Moses Whitbeck

and the
102nd NY
Volunteers
during the
War of the Rebellion

Chris Whitbeck

MOSES WHITBECK AND THE 102ND NY VOLUNTEERS
DURING THE WAR OF THE REBELLION

Cover photo of Moses Whitbeck, 102nd NY Volunteers

ISBN 979-8-218-35624-8

Manufactured in the United States of America
First printing January 2024

Contents

Foreword

I was inspired by a series of articles published by the Reno Gazette Journal in 1996 about the Donner Party and what happened to them. The articles appeared in the newspaper every day chronicling their trip with a post each day about where the Donner party was that day and what they were doing. As a person who loves American history, I was intrigued by the series and looked forward to each day's newspaper and article as I read and watched the Donner Party's Journey.

My father, Donald Whitbeck Jr., gave me the Union Army Discharge certificate for Moses Whitbeck, Sergeant, One hundred and second New York State Volunteer Infantry Regiment. My great uncle, Frank Whitbeck Jr., gave me a copy of the family history which included his letter (included) about his personal research into the life of Moses Whitbeck, his grandfather. After I read Uncle Frank's letter I was intrigued. I had always loved World War II history when I was in high school and until about 20 years ago when I started studying Civil War History. I started doing some research about the 102nd NY Volunteers and Discovered the War of the rebellion Series published by Congress starting in 1881. As I was researching, I thought this would make a wonderful story to not only discover, but to write about it and pass it down to my family.

I found books written by two commanders of the Divisions in the Corps that the 102nd was attached to. The Civil War letters of General Alpheus S. Williams are preserved in the book From the Canon's Mouth, by Wayne State University Press and the Detroit Historical Society. General Williams commanded the First Division and sometimes became the Corps commander when that commander was injured or disabled. The citations from this book will be listed as FTCM. The Civil War letters of General John W. Geary are preserved, and in the book A Politician Goes to War, by The Pennsylvania State University Press. General Geary was the Commander of the Division that the 102nd NY Volunteers were attached to for most of their service. Both of these books were invaluable aids in helping me find more information about the 102nd NY Volunteers and where they went and what they did. The Citations from General Geary's book will be listed as APGTW.

I am also grateful to the late R. L. Murray for his work about NY regiments in the Civil War and his book A Perfect Storm of Lead, which gave many insights and details about the NY regiments, including the 102nd NY Volunteers, in the Battle of Gettysburg and the defense of Culp's Hill. In the book the citations will be listed as APSL.

Of course, this research could not have been conducted without the help of the War of the Rebellion Series and also that Cornell University Press put them all online for anyone to use. If, you are not familiar with the series, the official citation is: MLA. United States. War Department. The War of the Rebellion: A Compilation of the Official Records of the Union and Confederate Armies. Some researchers refer to them as the O.R. In this book they will be referred to as WOTR.

I certainly used the O.R. and spent countless hours scouring them and putting together my book. When I use the term General Geary "wrote" that means the information came from his book A Politician Goes to War. When I use the term "reported" that means the information comes from the War of the Rebellion series. The same goes for information from

General Williams but comes from his Book From the Canon's Mouth.

In the summer of 2022, I took an epic trip to visit Antietam, Harpers Ferry, Gettysburg, and followed the path of the 102nd NY through 5 states. That brought to life for me the epic size of the Civil War and its impact on everyone. Antietam was especially moving, as I walked the Bloody Lane and discovered that Moses' unit was between the Bloody Corn field and the Bloody Lane, where at each location 6,000 casualties happened in about two hours fighting. I cried as I walked down the Bloody Lane, it was overwhelming. I spent several visits to Culps' Hill, in the morning and in the evening so I could get a feel for what the battle was like. I walked down to the stream where the Confederates were and looked back. You have to have guts to attempt to go up that while the Union was picking your men off, is what I thought at the time. Also, the Burnside Bridge at Antietam is incredible as it brings to life the odds the Union faced to cross it and how many men died trying. If you ever have a chance, go visit these places!

Although I have done much research, I have never found a diary from anyone in the 102nd NY Volunteers. I hope that by publishing this book someone will contact me with more information about the 102nd NY that is not in the public record. Also, maybe some long lost family relative with other possessions of Moses Whitbeck will be found. I can only hope as I research and write my next book about the 93rd Illinois that my other Great, Great Grandfather, Jonathan Miller fought with.

Either way, I hope you like the book and the style. I realize there may be parts where you are wishing for more information, Let me say that if I would have found it for the most part I would have added it. Some of the people and places there is very little information about. But, it's amazing just how much information there is, especially the bigger the battle, engagement, rank or personality. I am grateful to find what I did about Moses Whitbeck and the 102nd NY Volunteers. I am also grateful to my Great Uncle Frank Whitbeck for the research he did and inspiring me to finish the story. I also went to Shawnee, Oklahoma, and visited Moses' grave site. I felt overwhelmed again that he survived the Civil War and had two families as I am result from the Second family. Without God's protection and providence this book would have never been written.

Introduction

The 102nd New York Infantry was organized in New York City beginning November 1, 1861.

This is the official record of the 102nd NY Volunteers Infantry Regiment "Van Buren Light Infantry" below. It comes from the War of the Rebellion series. Officially titled "The War of the Rebellion: A Compilation of the Official Records of the Union and Confederate Armies," the Official Records are compiled in 127 volumes, plus a General Index and accompanying Atlas. The editor was Robert N. Scott. Citations from this Source will be listed as: WOTR with Volume and page noted.

Organized at New York City. Left New York State for Washington, D.C., March 10, 1862. Attached to Wadsworth's Command, Military District of Washington, to May 1862. Cooper's 1st Brigade, Sigel's Division, Dept. of the Shenandoah, to June 1862. 1st Brigade, 2nd Division, 2nd Army Corps, Pope's Army of Virginia, to August 1862. 2nd Brigade, 2nd Division, 2nd Army Corps, Army of Virginia, to September 1862. 2nd Brigade, 2nd Division, 12th Army Corps, Army of the Potomac, to October 1862. 3rd Brigade, 2nd Division, 12th Army Corps, Army Potomac, to October 1863, and Army of the Cumberland to April 1864. 3rd Brigade, 2nd Division, 20th Army Corps, Army of the Cumberland, to June 1865. 1st Brigade, Bartlett's Division, 22nd Army Corps, Dept. of Washington, to July 1865.

SERVICE: Duty in the Defenses of Washington, D.C., until May 1862. Moved to Harper's Ferry, W. Va., May. Defense of

Harper's Ferry against Jackson's attack May 28-30. Operations in the Shenandoah Valley until August. Battle of Cedar Mountain August 9. Pope's Campaign in Northern Virginia August 16-September 2. Guard trains during the campaign. Maryland Campaign September 6-22. Battle of Antietam September 16-17. Duty at Bolivar Heights until December. Reconnaissance to Rippon, W. Va., November 9. Expedition to Winchester December 2-6. March to Fredericksburg, Va., December 9-16.

At Fairfax Station until January 20, 1863. "Mud March" January 20-24. Regiment detached in New York on special duty March 10-April 4. Chancellorsville Campaign April 27-May 6. Battle of Chancellorsville May 1-5. Gettysburg (Pa.) Campaign June 11-July 24. Battle of Gettysburg July 1-3. Pursuit of Lee to Manassas Gap, Va., July 5-24. Duty on line of the Rappahannock until September. Movement to Bridgeport, Ala., September 24-October 3. Reopening Tennessee River October 26-29. Guarding railroad until November. Chattanooga-Ringgold Campaign November 23-27. Battles of Lookout Mountain November 23-24; Mission Ridge November 25; Ringgold Gap, Taylor's Ridge, November 27.

Duty in Lookout Valley until May 1864. Atlanta (Ga.) Campaign May 1-September 8. Demonstrations on Rocky Faced Ridge May 8-11. Battle of Resaca May 14-15. Near Cassville May 19. Advance on Dallas May 22-25. New Hope Church May 25. Battles about Dallas, New Hope Church and Allatoona Hills May 26-June 5. Operations about Marietta and against Kennesaw Mountain June 10-July 2. Pine Hill June 11-14. Lost Mountain June 15-17. Gilgal or Golgotha Church June 15. Muddy Creek June 17. Noyes Creek June 19. Kolb's Farm June 22. Assault on Kennesaw June 27. Ruff's Station, Smyrna

Camp Ground, July 4. Chattahoochee River July 5-17. Peach Tree Creek July 19-20. Siege of Atlanta July 22-August 25. Operations at Chattahoochee River Bridge August 26-September 2. Occupation of Atlanta September 2-November 15. Expedition from Atlanta to Tuckum's Cross Roads October 26-29. Near Atlanta November 9. March to the sea November 15-December 10. Near Davidsboro November 28. Siege of Savannah December 10-21.

Campaign of the Carolinas January to April 1865. Battle of Bentonville, N. C., March 19-21. Occupation of Goldsboro March 24. Advance on Raleigh April 9-13. Occupation of Raleigh April 14. Bennett's House April 26. Surrender of Johnston and his army. March to Washington, D.C., via Richmond, Va., April 29-May 20. Grand Review May 24. Duty at Washington, D.C., until July. Mustered out July 21, 1865.

Regiment lost during service 7 Officers and 66 Enlisted men killed and mortally wounded and 82 Enlisted men by disease. Total 155.

WHITBECK, MOSES.—Age, 21 years. Enlisted, November 13, 1861, at Rondout, to serve three years; mustered in as private, Co. G, November 15, 1861; promoted corporal, August 1, 1863; sergeant, January 1, 1864; re-enlisted as a veteran, January 30, 1864; returned to corporal, July 10, 1864; promoted sergeant, November 1, 1864; sergeant-major, July 1, 1865; mustered out with regiment, July 21, 1865, at Alexandria, Va.

New York State Military Museum official Unit roster

1862

January 27, 1862. The Van Buren Light Infantry had one company of Von Beck riflemen recruited out of New York City and another company of Von Beck riflemen recruited and assembled at Rondout, New York, beginning Nov. 1, 1861. Moses Whitbeck, a farmer in Ulster County, New York, had enlisted on Nov. 13, 1861, and was enrolled in Co. F, 1st Regiment, Von Beck Rifles. This organization became Company G of the 102nd N.Y. Volunteers on Jan. 27, 1862, with the consolidation of the Von Beck Rifles under Col. R. H. Shannon, and part of the McClellan infantry under Col. S. Levy, with Colonel Van Buren's command. (Side note: This author's birthday was Nov. 13, 1960. Is that a coincidence?)

Rondout is located on the Hudson River, near the mouth of Rondout Creek. The author went there in the summer of 2022. The area became an important hub for the shipment of timber and agriculture and the area grew when the Delaware and Hudson Canal opened in 1828. It became a trade hub for coal from northeast Pennsylvania, bluestone from the Catskill Mountains, cement from Rosendale, and bricks made from local clay. The area grew quickly, and in 1849 it was incorporated as the Village of Rondout and by 1872 it joined with the Town of Kingston.

Also, on this date in history, January 27, 1862, President Lincoln issued a war order authorizing the Union to launch a unified hard-hitting campaign against the Confederacy. General McClellan basically ignored the order. He believed his army wasn't ready and the shuffling of commands, as we see with the 102nd NY, may have been an indicator of that.

On **March 5, 1862,** the 102nd NY Volunteers received their regiment designation. They became part of the Van Buren Light Infantry. The men who enlisted in Rondout with Private

Moses Whitbeck were assigned to be Company G of the 102nd NY Volunteers.

The regiment at the time was led by Colonel. R. H. Shannon, and were part of the McClellan Infantry, led by Colonel. S. Levy. Now that the organization of the regiment was complete, the men of the 102nd NY were mustered in the service of the United States for three years starting September 9, 1861, and continued to gain more men all the way until April 5, 1862.

The organization was considered as completed in March 1862, although the men had enlisted at various times during the five preceding months. While organizing, the men were positioned at New Lots, Kings County, which is in Brooklyn. The field officers first commissioned were: Thomas B. Van Buren, Colonel; William B. Hay-ward, Lieutenant Colonel; and James C. Lane, Major.

March 10, 1862, the 102nd NY Volunteers Regiment, with eight companies and more than 1000 men strong, under command of Col. T. B. Van Buren, left the State of New York from Brooklyn heading to Washington to help defend the capitol and build its defenses. (Companies I and K left April 7, 1862). The Regiment served at and near Washington, D. C., in Abner Doubleday's Brigade, Wadsworth's Division, 1st Corps, Army of the Potomac.

At the outbreak of the war, Washington's only protection was an old fort, Fort Washington, 12 miles to the south, with a small force of Union regular army soldiers. It is interesting to note that there were less than 25,000 total men in the whole army before the start of the war. After Maj. Gen. George B. McClellan assumed command of the Department of the Potomac on August 17, 1861, he began to build a defensive network for the city of Washington. McClellan devised a plan for a complete circle around the city of Washington with entrenchments and fortifications. He built walled forts on high hills around the city and placed batteries of field artillery in the spaces between these fortifications. In between those, rifle pits were built, allowing co-operative fire that would make it hard for any force to overcome. That plan, once finished, made Washington D.C. one of the best defended cities in the world. The 102nd NY Volunteers helped build some of these defenses.

March 15, 1862, the 102nd NY Volunteers crossed the Potomac and marched to Langley, Va., where they were stationed on the outposts for ten days, after which they returned to Washington. The official records don't say much about this time, but we understand that the men were involved in helping build defenses around Washington and were probably sent to Langley to help build defenses there.

April 2, 1862. The 102nd NY Volunteers were not a professional army unit but came from various walks of life. The Company G from Rondout contained many farmers from Ulster County. The Regiment had been previously stationed in Washington to help build the Capitol's defenses. There they would also begin training to make them into an effective fighting unit.

Their commander, General Wadsworth, wrote, "In regard to the character and efficiency of the troops under my command, I have to state that nearly all the force is new and imperfectly disciplined, that several of the regiments are in a very disorganized conditions from various causes…and of little or no value in their present position…it is, in my judgement, entirely inadequate to, and unfit for, the important duty to which it is assigned." WOTR Vol 12 P. 225

April 7, 1862. The 102nd NY Volunteers are completed, as Companies I and K joined the regiment. The Regiment served at and near Washington, D. C., in Abner Doubleday's Brigade, Wadsworth's Division, 1st Corps, Army of the Potomac. At this point the regiment was helping Union General George McClellan to prepare defenses in Washington, D.C. Little do they know but soon they will be leaving for the defense of Harpers Ferry as Confederate General "Stonewall" Jackson is advancing up the Shenandoah Valley towards Washington, D.C.

April 19, 1862. The 102nd NY Volunteers were assembled at the Aqueduct, Washington, D.C., with other regiments for training and organization. The Washington Aqueduct, built by the Army Corps of Engineers between 1853 and 1863 was the District of Columbia's first public water system. It was a significant engineering achievement. It was designed by Army Engineer Montgomery C. Meigs, who also supervised the project until 1861, when he was appointed Quartermaster General of the U.S. Army. At that time the Army Corps of Engineers had an impressive reputation because of the Washington Aqueduct and their status in the field of public works reflects the military influence on life in America at that time in history. The Washington Aqueduct still supplies the nation's capital with public water today.

In **May 1862**, the regiment was assigned to General Abner Doubleday's Brigade, of Rufus King's Division, Irvin Mc-Dowell's First Corps, and was stationed at Aquia Creek, which is south of Washington, not far from Fredericksburg, Virginia.

Abner Doubleday initially served in coastal forts and then in the Mexican-American War from 1846 to 1848 and the

Seminole Wars from 1856 to 1858. In 1858, he was transferred to Fort Moultrie in Charleston Harbor serving under Colonel John L. Gardner. At the start of the War of the Rebellion, he was a captain and second in command at the garrison at Fort Sumter, under Major Robert Anderson. He aimed the cannon that fired the first return shot in answer to the Confederate bombardment on April 12, 1861. He subsequently referred to himself as the "Hero of Sumter" for his role.

Rufus King was appointed by President Abraham Lincoln as Minister to the Italian Papal States in 1861 after being nominated by Secretary of State William H. Seward. The Civil War broke out when King was on his way to Italy so he took a leave of absence to join the Union Army. He was appointed a Brigadier General of the Wisconsin militia on April 15, 1861, and of U.S. Volunteers on May 17, and was granted permission to raise a Wisconsin regiment. King helped organize and lead what came to be known as the famous Iron Brigade, which was composed of the Second, Fifth, Sixth, and Seventh Wisconsin, and Nineteenth Indiana volunteers. Commanding the 1st Corps was career Army officer Irvin McDowell. He is best known by his defeat in the First Battle of Bull Run, which was the first large-scale battle of the war. In 1862, he became Commander of the 1st Corps of the Army of the Potomac. He fought unsuccessfully against P.G.T. Beauregard's command during the Shenandoah Valley Campaign of 1862 and was blamed for contributing to the defeat of United States troops at the Second Battle of Bull Run in August. Wikipedia: Irvin McDowell

Defense of Harpers Ferry West Virginia against Jackson's attack, May 28-30, 1862.

May 29-30, 1862 — As part of the Shenandoah Valley

Campaign of 1862, Confederate General "Stonewall" Jackson's forces attacked Harpers Ferry, West Virginia, on **May 30, 1862**. Because of the town's strategic location at the corner of West Virginia, Maryland and Virginia where the Potomac River meets the Shenandoah River, and where the Baltimore & Ohio Railroad runs through, both Union and Confederate troops frequently moved through Harpers Ferry. Harpers Ferry was difficult to defend as it was surrounded on all sides by steep rises leading to Bolivar Heights, Loudoun Heights and Maryland Heights. The lower elevation of the town made it a sitting duck for any artillery posted above. The town did not have a good reputation with Soldiers, being called a "godforsaken, stinking hole." Battlefield.org: 10 Facts Harpers Ferry

During the campaign, General Jackson defeated all of the Union troops in the Shenandoah Valley, except those at Harpers Ferry. Because of Jackson's achievements in the Shenandoah Valley Campaign, General Rufus B. Saxton and his men were sent to reinforce Harpers Ferry. This lead to the Second Battle of Bolivar Heights. Jackson's troops approached Harpers Ferry from Bolivar Heights and charged to within a few hundred yards of Union fortifications on Camp Hill during a ferocious thunderstorm.

The Union had cannons on Maryland Heights and they struck Jackson's men as they advanced across the heights through the town of Bolivar, damaging numerous homes. Saxton's position on Camp Hill was held, and Jackson withdrew that night. (In 1093, Saxton would be awarded the Congressional Medal of Honor for his actions that day.) The 102nd NY Volunteers and Private Moses Whitbeck were under the command of Brigadier General Cooper in the 1st Division of the 1st Corps. They were stationed on Maryland

Heights and were not directly engaged in this battle by Confederate troops.

June 4, 1862. The 102nd NY Volunteers arrived in Winchester, Virginia. Located in the Shenandoah Valley, Winchester was the most contested town in the Confederacy during the American Civil War, changing hands several times and earning its reputation (in the words of a British observer) as the shuttlecock of the Confederacy. When war broke out, Winchester's location and function as a rural market center ensured that it would be wanted by both sides in the conflict. Possessing Winchester would be necessary to control the Shenandoah Valley's abundant agricultural resources. Further, possession of Winchester had broad strategic implications. A Confederate army in Winchester would be north of Washington, D.C., and could threaten the capital or open the way to an invasion of Maryland or Pennsylvania. A Union army in Winchester, meanwhile, could jeopardize Confederate Gen. Robert E. Lee's extended left flank and his ability to protect the Confederate capital at Richmond. Winchester was viewed by many as the key that locked the door to Richmond. As much as Winchester was a prized target, it proved especially difficult to keep. The town was surrounded by low hills that easily concealed approaching armies, and neither side was successful in holding it against an approaching army. Encyclopedia Virginia.org : Winchester during the Civil War

June 8, 1862. The 102nd NY Volunteers were still in Winchester getting organized and trained. Winchester was relevant to several significant military operations during the course of the war. The town was the site of an important Confederate victory on May 25, 1862. This First Battle of Winchester was part of Stonewall Jackson's Shenandoah Valley Campaign, which served to distract Union troops from reinforcing George B. McClellan outside of Richmond.

While local historians claim that Winchester changed hands more than seventy times during the war, estimations of full-fledged occupations by either army range from eleven to thirteen. Wartime diaries suggest that Winchester was under Confederate authority for 39 percent of the war, occupied by Union armies for 41 percent of the war, and between the lines for 20 percent of the war. As a result of continued defeat in the Shenandoah Valley and evolving military policy, each successive Union occupation resulted in harsher measures toward civilians. Initially individuals were harassed and home-steads pillaged. In spring 1862, Union Gen. Nathaniel P. Banks attempted to appease Winchester's population. Union Gen. Robert H. Milroy, however, admitted he felt "a strong dis-position to play the tyrant among these traitors," embittering residents with his harsh policies throughout the first half of 1863. He required citizens to take oaths pledging their al-legiance to the United States. If they refused, soldiers would quarter their homes. Milroy also permitted Union troops to obliterate Winchester, refusing to interfere when they de-stroyed every unoccupied house in town. Encyclopedia Virginia.org : Winchester during the Civil War

June 26, 1862. Maj. Gen. John Pope was placed in com-mand of the newly constituted Army of Virginia. The 102nd NY Volunteers, commanded by Colonel Joseph C. Lane, were part of the Union Army of Virginia II Corps under General Banks, in the Second Division under General Augur, in the 2nd Brigade under Brigadier General Henry Prince. Pope's orders were to defend Washington DC and Union-held northern Virginia while the Army of the Potomac under Maj. Gen.

George B. McClellan battled Robert E. Lee outside of Richmond. When McClellan was defeated at the end of the Seven Days battles less than a week later, Lee turned his attention north toward Pope while McClellan regrouped his army. Pope's three army corps were arrayed in a line from the Blue Ridge Mountains to the Rappahannock River. Lee responded to Pope's dispositions by dispatching Maj. Gen. Thomas J. "Stonewall" Jackson with 14,000 men to Gordonsville toward the center of Pope's line. Jackson was later re-inforced by Maj. Gen. A.P. Hill's division.

On **August 6, 1862**, General Pope marched his forces south into Culpeper County with the objective of capturing the rail junction at Gordonsville, Virginia. Recognizing its importance, the Union army made several failed attempts to capture Gordonsville during the war and thereby cut off the Confederate military and supply line. In addition to Gordonsville's role on the Confederate supply line, it also contained an important medical hospital. In March 1862, the town's premier inn, the Exchange Hotel, was converted into a battlefield hospital.

Gordonsville and the railroads that crossed there were of vital importance to the Confederacy for troop mobility and supplies. Troops from Richmond going to the First Battle of Bull Run on July 21, 1861, came through town. During the war years, James Longstreet, Stonewall Jackson, Richard S. Ewell, A. P. Hill, and Robert E. Lee spent time in Gordonsville. Major Gen. Philip Sheridan led a raid toward Gordonsville and Charlottesville but was stopped by Wade Hampton's Confederate cavalry in the vicinity of Trevilian Station. Gordonsville was threatened many times but was always successfully defended by the Confederates. During its four-year use, the medical center treated more than 70,000 Confederate and captured Union soldiers. The Civil War ended in 1865 and with Gordonsville being largely unscathed, passenger rail service was quickly reestablished. During Reconstruction (1865–1877), the Freedman's Bureau used the hotel as a hospital for newly freed African Americans. Today the restored late-Greek Revival–style Exchange Hotel serves as a Civil War museum. Encyclopedia Virginia.org : Gordonsville during the Civil War

The Battle of Cedar Mountain, August 9, 1862

Confederate Gen. Jackson and Union Maj. Gen. Nathaniel Banks's Second Corps of Pope's Union Army of Virginia tangled at Cedar Mountain, Virginia. There was some confusion among the leaders of the Confederate forces, and before leadership could properly be restored to the division, the Union attack involving the II Corps Brigades of Geary and Prince (including the 102nd NY Volunteers) were sent against the Confederate right.

The 102nd NY Volunteers marched in the morning from Culpepper Courthouse following the First Brigade and took the main road southward toward Cedar Run, Virginia. The men could hear booming canons indicating they were not far from the battle of Cedar Mountain. It was an extremely hot day, and about six miles from Culpepper the men found a shady wooded area where they rested and drank water. The men were then informed to get ready for battle. They passed through the woods and filing to their left followed a small creek about three fourths of a mile and then crossed it and awaited orders. They were then ordered to form a line of battle and march to the left of General Geary's brigade. The 102nd was placed behind the 111th Pennsylvania and 3rd Maryland

Regiments. The brigade advanced under heavy fire and because of the terrain the 3rd Maryland fell a little behind the 111th Pennsylvania. The Union infantry continued to sweep forward against the enemy, their "huzzahs" screamed against the roar of musket and cannon fire. General Prince moved the 111th Pennsylvania and 3rd Maryland through the corn toward Confederate General Early, then received orders to throw his entire brigade at the Confederates en echelon. The orders, executed under extreme duress, were dispatched with a notation for the 109th Pennsylvania and 102nd New York to take care not to fire into the lead regiments. Even though they tried to be extremely careful, just such a volley was fired. When the 102nd NY fired their first volley, the 3rd Maryland thought they were being attacked from behind and retreated in disorder. This caused the 111th Pennsylvania to be exposed by themselves, and they too retreated. The 109th Pennsylvania and 102nd NY regiments moved forward without delay to fill the gap. The 109th volleyed into the 12th Georgia, while the 102nd wheeled right and opened fire on Thomas' 52nd Virginia recently arrived brigade. There was some confusion on the Confederate side, and this allowed the Union forces to advance and threaten to break the Confederate lines. Confederate General Early was able to rally his troops, and the rebels raked fire into the union lines. The 102nd NY was on the Union left and repulsed by the 52nd Virginia and the 12th Georgia regiments. Early's stabilizing presence and the raking fire of the Confederate guns halted the Union advance on the Confederate right.

A Confederate counterattack led by General A.P. Hill on the Union right, which included Stonewall Jackson's brigade coming from the rear in support at the eleventh hour, repulsed the Federals and won the day. As the leftmost unit during the Union advance on the southern portion of the battlefield, the 102nd New York was exposed to Confederate artillery fire. General Prince, who commanded the Second Division, wrote: "The conduct of the brigade, considering its advanced position and severe combat, was highly credible to it. This will be fully appreciated by the table of casualties showing a loss of 33 percent." WOTR Vol 12 pt. 2 p. 169

In the confusion of the 3rd Maryland and 111th Pennsylvania retreat, General Prince rode to the rear trying to rally the troops and was captured by Confederate forces. He would be released late that summer in a prisoner exchange, and it was after that he filed his official report.

The 102nd NY Regiment was commanded by Major Lane. It lost Captains Julius Spring and Arthur Cavanaugh and 21 enlisted men killed or mortally wounded, Lieutenant Colonel Avery and 6 other officers and 70 enlisted men wounded, and 1 officer and 14 men missing. Confederate Brig. Gen. Charles Winder was killed. General Geary received a wound to his elbow and would be sent home for several weeks of recuperation. Private Moses Whitbeck survived the battle. The battle at Cedar Mountain shifted fighting in Virginia from the Peninsula to Northern Virginia, giving Lee the strategic initiative. Excerpts taken from American Battlefield Trust: Cedar Mountain

(Note: there is no report in the WOTR from Major Lane who commanded the 102nd NY Volunteers)

August 12, 1862. Pope's Army makes it to Culpeper Court House. They are battered and bruised as Princes' division, which contains what remains of the 102nd NY Volunteers. Culpeper itself would be a focal point for military action. Geographically, it sat midway between and slightly to the west of Richmond and Washington, D.C., and railroads linked it to both national capitals. The Orange and Alexandria ran northward from the county seat of Culpeper Court House to Alexandria; the Virginia Central connected the county to Richmond via Gordonsville. In addition, the Rappahannock River formed the county's northern boundary, and Culpeper marked the first point on the river where an invading Union force could ford the Rappahannock during most of the year. Outside of the Shenandoah Valley, it was one of the best invasion routes in the state. Encyclopedia Virginia.org: Culpeper County during the Civil War

August 16, 1862. Because of the health of the Corps as they recover from the battle, the regiment is assigned to guard trains (wagons, trains, supply depot) during Pope's Campaign in Northern Virginia August 16-September 2. Trying to find information about this is difficult as most of the reports of this time involve major movements to or from, or during, battles.

Second Battle of Bull Run August 28-31, 1862. The 102nd NY Volunteers were in reserve, Guarding Harper's ferry and missed the battle.

The Battle of Chantilly (or **Ox Hill,** the Confederate name) took place on **September 1, 1862,** in Fairfax County, Virginia, as the concluding battle of the Northern Virginia Campaign in 1862. Thomas J. "Stonewall" Jackson's corps of the Army of Northern Virginia attempted to cut off the line of retreat of the Union Army of Virginia following the Second Battle of Bull Run but was attacked by two Union divisions. During the ensuing battle, Union division commanders Isaac Stevens and Philip Kearny were both killed, but the Union attack halted Jackson's advance. The 102nd NY was still on guard duty up until this time.

Maryland Campaign September 1–22, 1862.

September 1, 1862. The 102nd NY Volunteers under The 12th Corps Second Division, left Bull Run and marched toward Fairfax, camping at a fork in the road General Geary reported. Following the Confederate retreat from Antietam, the Union re-occupied Fairfax (sometimes called Fairfax Courthouse) through 1862 and the first half of 1863. First the XI Corps and then a brigade of Vermont regiments set up camps. The white civilian population, by this time, was showing signs of the privation the war brought to them. The armies had picked over their storerooms for food and had even stripped their homes for firewood after taking their fences. Many civilians tried to make a living by cooking or cleaning for the thousands of Union troops now camped in town. Since there was no market, no fences to restrain livestock, and no crops that could be grown without being procured by the army, citizens had little alternative than to work in support of the occupying military forces.

These camps that were made near small cities were often virtual tent cities themselves, inhabited by soldiers from both armies during their stay, as was this one in Fairfax. Organized to keep men from the same company and regiment sharing space, these large rectangular camps included tents for officers and enlisted men, sinks (toilets), guard posts, kitchens, trash heaps, hospitals, sutlers (camp stores), stables, and stockpiles of supplies. They functioned much like a small city. In an era when there were only nine cities with populations over 100,000, and only 46 with populations over 20,000, the presence of a camp of 50,000 men instantly became one of the most densely populated communities of people in the whole country. The armies had to supply both military and domestic goods to provide for the sustenance, comfort, and fighting readiness of the troops. This meant that the camps in Fairfax were important points in the military networks that had to be constructed to keep armies in the field. The Virginia Department of Historic Resources: Fairfax Court House 1861-1865

September 2, 1862. The 12th Corps marched toward Alexandria, halting near Fort Worth. In the spring of 1862, Alexandria had become the "Crossroads of the Civil War." The Confederate victory in the Battle of Bull Run 10 months earlier sent wounded soldiers from both sides of the conflict flooding into the city, presenting an unprecedented and overwhelming scale of casualties that brought a desperate need for more hospitals. Mansion House, which became the largest of these hospitals at more than 500 beds, was confiscated from prominent Alexandrian James Green, and had been the most luxurious hotel in the region.

The once prosperous Southern port and railroad town was now a melting pot including military personnel and civilians on both sides of the conflict – nurses, doctors, wounded soldiers and a growing population of escaped slaves who were flooding the city to achieve relative freedom and employment opportunities. Technically still the property of their former masters, these escapees were labeled "contraband" in order to prevent them from having to be returned. By 1863, some estimates put the number of contrabands in the city at 18,000, an increase of more than 10,000 in a period of just 16 months.

Fort Worth was constructed as a timber and earthwork fortification west of Alexandria as part of the defenses of Washington, D.C. It was built after the Union's defeat at Bull Run and overlooked the Orange and Alexandria Railroad, the Little River Turnpike, and the southern approaches to the city of Alexandria. It was also the largest settlement in Union-occupied Northern Virginia.

September 3, 1862, The Second Division of the Union Army 12th Corps, marched beyond Alexandria, and halted in the rear of Fort Richardson. Fort Richardson was a detached redoubt that the Union Army constructed in September 1861 as part of the Civil War defenses of Washington. The Army built the fort shortly after its rout at the First Battle of Bull Run (Manassas) in late July 1861. The Army named the fort after General Israel B. Richardson, whose division had been deployed to defend the City of Washington against attack by way of the Columbia Turnpike. The fort, which was the highest fortification on the Arlington Line, occupied an imposing position on the crest of a ridge. It had a perimeter of 316 yards and emplacements for 15 guns, including a 100-pound Parrott rifle that could sweep a sector from Fort Ellsworth to Fort DeKalb (later named Fort Strong). The fort housed "bomb-proofs" and two ammunition magazines and was adjacent to a military encampment.

September 4, 1862. The Union Army 12th Corps marched through Georgetown and camped near Tennallytown. Georgetown had a college and buildings on the south side of campus, as well as the College's villa in Tenleytown, (alternate spelling used during the war) were requisitioned because of a pressing need for Hospital accommodation for 500 patients. The "new building" mentioned appears to be Maguire Hall, construction on which began in 1854. Apparently, "Old North" as it was called, was spared because of the intercession of a Union General, General Amiel Weeks Whipple, whose sons William and David were enrolled there. Because of this, Georgetown was able to continue in operation, although enrollment dropped to as low as seventeen students. The campus was not returned to the college's control until February 1863. Tennallytown was the site of several forts on the outskirts of Washington. Shades of Blue and Gray: Georgetown and the Civil War | Georgetown University Library. https://library.georgetown.edu/exhibition/shades-blue-and-gray-georgetown-and-civil-war

Directing Maj. Gen. Samuel P. Heintzelman to guard the capital with the III and XI Corps, Union General McClellan began marching north on **September 4** with about 84,000 men. He formed the six remaining corps into three subordinate commands and a reserve. Burnside commanded the right wing, which consisted of the I Corps and his own IX Corps; Maj. Gen. Edwin V. Sumner—the oldest active corps commander in the Civil War at age sixty five—commanded his own II Corps and the 12th Corps in the Center Wing; and Maj. Gen. William B. Franklin's VI Corps, reinforced by an additional division, constituted the left wing. Maj. Gen. Fitz-John Porter followed with the V Corps as a reserve.

September 5, 1862. The Second Division marched through Tennallytown and camped near Rockville. Rockville in 1860 was a thriving center of commerce and local government. Although small, the town was the intersection of several major roads, making it strategic to the movement of troops and supplies during the war. By spring of 1861, some 10,000 Federal troops were stationed nearby, most of them camped on the Fairgrounds, now the location of Richard Montgomery High School. 1862 saw the Courthouse used as a field hospital after the bloody Battle of Antietam and the beginning of arrests of local pro-South citizens for "disloyalty."

September 6, 1862. The Union Army 12th Corps moved up and took position in line, which had been formed about 2 ½ miles from Rockville. There was much movement by Union forces, especially Calvary and probing to determine just where Lee's army was and what its strength was. The 12th Corps would remain here for a few days between Rockville and Middlebrook. There was a lot of caution by Union General Meade as he overestimated Lee's forces to be about 120,000 men. Also Lee's intentions were not exactly clear at the time. It was thought that Lee was moving on Baltimore, and the Union forces faced an unordinary number of stragglers, so much so that General McClellan had to deal with it by issuing strict orders against it.

By **September 7, 1862**, as the last Confederates crossed the Potomac to join the rest of Lee's army at Frederick, Union advanced units reached Rockville, Maryland, just twenty-five miles away.

September 9, 1862. The 12th Corps Marched to Middlebrook. Union General Sumner's column took the direct route toward Frederick and reached Middlebrook. McClellan then held a sixteen-mile front just east of Parr's Ridge, halfway between Washington and Frederick, and much closer to the Army of Northern Virginia than Lee realized. The stage had been set for a great confrontation.

September 10, 1862. The 12th corps marched toward Damascus as part of an overall movement by Union forces and camped within two miles of there. Early in the morning while the different Corps were on the march to varying destinations, Gen. McClellan received information that the entire Confederate army was still at Frederick, and he wanted to verify this information by further reconnaissance. He would not press his advance until satisfied whether the enemy intended to move toward Baltimore or Washington. The whole Confederate army was in fact, on its "rollicking march" from

Frederick to Harper's Ferry and Hagerstown. The Confederates began leaving Frederick that day. Jackson's divisions initially marched northwestward, as if heading toward Pennsylvania, to mask the army's true intentions, then turned west to Williamsport.

September 11, 1862. The 12th Corps, which included the 102nd NY Volunteers, moved on to Damascus. In the afternoon, Jackson's command recrossed the Potomac into Virginia. As Jackson approached Martinsburg, Union Brig. Gen. Julius White ordered his 2,500 troops to evacuate to Harpers Ferry, fifteen miles away, where he, despite his rank, subordinated himself to Col. Dixon S. Miles. As the Confederates converged, Colonel Miles did not evacuate as Lee expected but prepared to hold the town. By evening, Confederate General McLaws had gained Maryland Heights from the northeast, Walker occupied Loudoun Heights to the south, and Jackson approached Bolivar Heights from the west, thus surrounding the Union garrison.

September 12, 1862. The 12th Corps took up a line of march toward Frederick, gaining 7 miles and then camped. The 102nd NY Volunteers were led by Lt. Col. James C. Lane and were in the Second Brigade led by Colonel Henry J. Stainrook, of the Second Division led by Brigadier General George S. Greene. U.S. Federal troops marched through the "village" of Damascus (via what is now Route 27) on their way to the town of Sharpsburg, where they engaged Confederate troops commanded by General Robert E. Lee at the Battle of Antietam.

Lee then traveled to Hagerstown with Longstreet's command on the 12th, while D. H. Hill's five brigades halted at Boonsboro to act as rear guard and to watch, as noted in

Lee's orders, "all the roads leading from Harper's Ferry, to intercept any Union forces that might escape."

September 13, 1862. The Union Army 12 Corps crossed the Monocacy River and camped near Frederick. When elements of the 12th Corps arrived at Frederick, Cpl. Barton W. Mitchell and Sgt. John M. Bloss of the 27th Indiana Infantry found an envelope containing a copy of Special Orders 191 wrapped around several cigars meant for General D. H. Hill. Sergeant Bloss realized that the document detailed Lee's plan of attack "and what each division of his army was to do." The two noncommissioned officers took it to their regimental commander, Col. Silas Cosgrove, who immediately carried it to Brig. Gen. George H. Gordon, their brigade commander. Gordon said what they found "was worth a Mint of Money & sent it to General McClellan."

The intelligence gained from the document disclosed that Lee intended to operate against Harpers Ferry and Martinsburg but ruled out Halleck's fear that the foray into Maryland was meant only to draw the Army of the Potomac north and leave Washington vulnerable to a rapid strike from the southwest. Discovery of the orders, at least temporarily, raised McClellan's confidence. "I have all the plans of the rebels," he informed Lincoln, "and will catch them in their own trap if my men are equal to the emergency…Will send you trophies." McClellan then boasted to Brig. Gen. John Gibbon, a brigade commander in the I Corps, "Here is a paper with which if I cannot whip 'Bobbie Lee,' I will be willing to go home." Telegram from George B. McClellan to Abraham Lincoln (September 13, 1862) - Encyclopedia Virginia: telegram-from-george-b-mcclellan-to-abraham-lincoln-september-13-1862

Still, McClellan found room for uncertainty. The captured orders were now four days old and may have been canceled, modified or superseded. Moreover, the orders disclosed nothing about the strength of the opposing army. McClellan continued to overestimate, believing that he faced "not less than 120,000 men." McClellan squandered the opportunity. His initial jubilation was overtaken by his caution. He believed that Lee possessed a far greater number of troops than the Confederates actually had, despite the fact that the Maryland invasion resulted in a high rate of desertion among the Southerners. McClellan was also excruciatingly slow to respond to the information in the so-called Lost Order. He took 18 hours to set his army in motion, marching toward Turner's Gap and Crampton's Gap in South Mountain, a 50-mile long ridge that was part of the Blue Ridge Mountains. History.com/this-day-in-history/union-troops-discover-rebels-antietam-battle-plan

Lee, who was alerted to the approaching Federals, sent troops to plug the gaps, allowing him time to gather his scattered units.

September 14, 1862. The 12th Corps marched toward South Mountain, camping near there. To reach Lee, McClellan needed to cross a major barrier, South Mountain, which was the northern extension of Virginia's Blue Ridge Mountains. The South Mountain range, some 70 miles long with elevations exceeding 2,000 feet, could only be crossed in a few places. Three of these passes—Turner's, Fox's, and Crampton's Gaps from north to south—thus became McClellan's immediate objective. During the morning, Maj. Gen. Jesse L. Reno's IX Corps of Burnside's wing moved west from Middletown and began reconnoitering Fox's Gap, where the Old Sharpsburg Road crossed South Mountain. Advance elements of the Army of the Potomac attempting to clear Turner's Gap encountered part of D. H. Hill's division. A confusing and sharp action developed in the rough mountain terrain of South Mountain, and although Longstreet reinforced Hill in the afternoon, by nightfall the Union First and Ninth Corps had captured the key positions that rendered the Confederates' position untenable. That night as Lee assessed the situation, he decided his campaign had failed and made plans to withdraw Longstreet and D. H. Hill to Virginia by way of Sharpsburg and to break off the Harper's Ferry operation, which he knew still had not concluded. Excerpts taken from the American Battlefield Trust. The Maryland Campaign of 1862

September 15, 1862. The Union Army 12th Corps, passed through Boonsborough and halted near Sharpsburg and camped there. Confederate General McLaws ignored Lee's orders to retreat to Virginia and during the early morning hours of September 15 assembled a defensive line across Pleasant Valley, facing Franklin's Sixth Corps. Even before he received Lee's orders to break off the operations against Harper's Ferry, Jackson had sent a courier speeding through the night with news that the surrender of the Union garrison was imminent. When Lee received Jackson's report on the morning of the 15th, he had already modified his plan to withdraw Longstreet and D. H. Hill to Virginia immediately, and ordered both commands to halt behind Antietam Creek, near Sharpsburg. There he would wait to see how McClellan responded to his success at South Mountain. If he pressed hard, Longstreet and Hill could still withdraw to Virginia. But if he moved cautiously, Lee might be able to concentrate his army and offer battle in Maryland. The reasons and the wisdom of Lee's decision to try and make a stand in Maryland have been debated ever since. He never fully explained his reasons, but by remaining in Maryland and hazarding a battle, he kept alive the possibilities the Maryland invasion had promised. If he could check McClellan and the Federals withdrew, then operations could be maintained on the Confederate frontier north of the Potomac and pressure continued on the North. If he withdrew to Virginia those opportunities were lost.

Meanwhile, the early morning of **September 15** at Harpers Ferry opened with a heavy artillery bombardment. By 8 a.m., the Union garrison of some 13,000 men, out of long-range ammunition for their artillery, were forced to surrender. At a cost of perhaps 400 casualties Stonewall Jackson won one of the most complete victories of the war for the Confederates. Meanwhile, McClellan learned that Lee had retreated from South Mountain, and reports from the front led him to believe that the rebels had been badly thrashed and were in rapid retreat for Virginia. He commenced an immediate pursuit of what he presumed was a fleeing enemy. But his euphoria was dashed that afternoon when reports came in of an enemy line of battle forming behind Antietam Creek, and Franklin reported a strong Confederate force standing defiantly across the approach to Harper's Ferry. Excerpts taken from the American Battlefield Trust. The Maryland Campaign of 1862

September 16, 1862. That night the Union Army 12th Corps moved up and took position on the left of General Mansfield's Corps, on the right of the line of battle. McClellan spent the morning waiting for a fog to lift, and then when it did, he busied himself with reconnaissance of the Confederate positions. While the hours ticked by, Jackson's two divisions and Walker's Division joined Lee at Sharpsburg. Straggling had reduced their strength by thousands, but they raised Lee's manpower to about 21,000. McClellan had about 60,000 effectives on hand. The odds were still long, and Lee sent urgent messages to McLaws to hurry to Sharpsburg. The Georgian, whose men were desperately short of rations, marched through the night, losing perhaps one-third of his strength to straggling, but he reached Sharpsburg before daylight on **September 17**. With McLaws and Anderson up, Lee had about 35,000 men. His army held a strong position with one flank anchored on the Potomac and the other on Antietam Creek. Excerpts taken from the American Battlefield Trust. The Maryland Campaign of 1862

Lee knew he would be attacked on the 17th. On the afternoon of **September 16**, McClellan at last set his army in motion, sending Major General Joseph Hooker's 1st Corps across Antietam Creek to find Lee's left flank. Just at dusk Hooker bumped into Hood's division and the two forces skirmished until dark. The night was pitch dark with a slight drizzle. "I shall not, however, soon forget that night," wrote Union General Alpheus Williams, "so dark, so obscure, so mysterious, so uncertain." FTCM p. 125

September 17, 1862. The 102nd NY Volunteers were commanded by Lt. Col. James C. Lane and part of the Second Brigade commanded by Col. Henry J. Stainrook, in Gen. George S. Greene's Second Division of the 12th Corps. The brigade was formed in a line of battle, in the face of the rebels, and already under fire when the order to charge was made. Stainrook's Brigade was on the side left of the Division, advancing in line south of the Smoketown Road, and relieved the left of Crawford's Brigade and engaged the Confederate Infantry in the East Woods. They drove the rebels back over a half of a mile. Stainrook's Brigade pursued through the East Woods, crossed the fields to the left of the burned out buildings of the Mumma farm, and halted behind the ridge a

Picture is of the Dunker Church as it stands today. The Dunkers were pacifists, and it is ironic that the war came right to them. The 102nd was just to the right of the picture.

few yards east of this point where with the assistance of Monroe's and Tompkins' Rhode Island Batteries, it protected the right of French's Division of the Second Corps, and repulsed several assaults of the enemy. After the union batteries were brought forward to the line gained, the right and center of the brigade rose from behind the battery and again drove the rebels back another 500 yards through another piece of woods, with great carnage. The regiment on the left was ordered by Gen. Sumner to remain behind the battery for support. About 10:30 a.m., the brigade crossed the road and entered the woods on the left of the Dunkard Church, its left on the road directly opposite, where it remained until noon when it was forced to retreat to the East Woods, from lack of support and want of ammunition, and at 1 p.m. was moved up in line about half a mile to rear of the line of battle and allowed to rest. At nightfall, the command was again ordered under arms and took position behind General Franklin's corps as a reserve and slept on their arms. Maj. Gen. Joseph F. K. Mansfield, who led the Corps, was mortally wounded, and Brig. Gen. Alpheus Williams temporarily took command.

From Lieutenant Colonel Lane's Official Reports of the 102nd New York at Antietam (War of the Rebellion Report No. 186 Volume XIX:

"I have the honor to report that in the action of yesterday September 17, 1862 the One hundred and second New York State Volunteers entered the field for duty, according to orders, at 6.30 a. m., in common with the rest of the brigade; that we marched to the woods held by the rebels in close column by division, and that line of battle was formed by deployment of column. While the line was forming, under fire of sharp-shooters of the enemy, Captain M. Eugene Cornell, of Company D of this regiment, fell, dead, at the front of his command while bringing them into line, being shot through the head. After line was formed we advanced in order, driving the rebel before us, this regiment going, however, to the left of the brigade, and, after passing through the woods, taking the left of the burning building in the field beyond. From this building our men pursued the enemy to the corn-field in advance, where the One hundred and second halted and commenced firing at a battery which was playing on the right of the brigade, just beyond the cornfield. This battery retired immediately after our opening fire upon it.

"At this time I marched the regiment by the right flank to rejoin the brigade, which was in position behind the battery of

Parrott guns, to the right of the corn-field. Soon after the brigade moved forward past the battery and drove the enemy through the woods beyond. The One hundred and second, however, remained supporting the battery, by order of General Sumner's aide. This battery retired after expending its ammunition, and was replaced by a battery of brass guns, which remained in position about twenty minutes, and returned, being threatened by a brigade of the enemy, the right of this brigade being out of ammunition and unsupported, retiring at the same time. The One hundred and second also retired, joining in with the rest of the brigade, and were reformed into line by the general commanding division (General Greene), at the rear of the woods behind the burning building. The brigade was here rested, and, after some two hours, was again marched one-half mile to the rear, and, after forming line, arms were stacked and ration given out." WOTH Vol 19 P. 510-511

Gen. Greene, commanding the Second Division reported, "Where so many acted with distinguished gallantry it is impossible to designate all the individual officers entitled to notice. Colonel Stainrook, lieutenant Colonel Lane, one hundred and second New York, and Major Pardee deserve commendation for activity and gallantry through the day." WOTH Vol 19 P. 505

The 102nd NY Regiment lost Cap. Eugene Cornell and seven enlisted men killed or mortally wounded, 24 men wounded, and five men missing. This was a lot less than other commands lost that day, as the Battle of Antietam was the bloodiest single day in American History. More than 23,000 casualties were suffered that one day by the Union and Confederate Armies combined. Pvt. Moses Whitbeck survived.

September 19, 1862. The Second Division marched in the direction of Harpers Ferry, being on the road all night, and arrived near Sandy Hook, Md., at 3 p.m. the next day on **September 20**. Sandy Hook, Maryland, is situated on the Potomac River a mile east of Harper's Ferry, West Virginia. Major General Nathaniel P. Banks established his army's headquarters at Sandy Hook when he assumed command of the Department of Shenandoah in late July 1861.

September 22, 1862. The Union Army 12th Corps took position on Loudoun Heights, Va., where they stayed until the end of the month. Loudoun Heights is the second highest mountain overlooking Harpers Ferry, hugging the Shenandoah River along its base. The north face of the mountain is a 900-foot vertical drop – the steepest bluff surrounding Harpers Ferry. It is named after Loudoun County, Virginia. Following the Battle of Antietam and the Union reoccupation of Harpers Ferry in late September 1862, Loudoun Heights became an extensive temporary campground for a portion of the Army of the Potomac. The place was miserable for camping, with limited water, steep slopes, boulder-strewn crests, and primitive roads. Remnants of these campgrounds still exist but are inaccessible due to the rugged terrain.

During the fall occupation of 1862, the Union army denuded the upper third of Loudoun Heights, making its crest bald and forlorn in appearance. By early November, the Federals had abandoned the mountain, and it never again played a significant role, as it fell under the big U.S. guns on Maryland Heights. Harpers Ferry National Historical Park (U.S. National Park Service): Loudoun Heights

September 30, 1862. The 102nd NY Volunteers were at Loudoun Heights, Virginia, near Harper's Ferry. General

Geary returned from his elbow wound he received at Cedar Mountain. General Geary wrote to his wife, "My command is upon these heights, and should we be attacked by the enemy, I am nearest him, and of course will have to engage him first, and perhaps stand the brunt of the action, well this has always been my luck, and it is mean compliment either to be considered the right kind of stuff for that place. All is going on well, and we are busy fortifying to the utmost, determined to make the most vigorous resistance should we be attacked; but what the rebels are about we cannot exactly tell. Things will be developed in a few days. My Old Veteran Brigade received me with great enthusiasm, and I find myself in command of a division. God grant that I may guide them to victory and success, under the direction of the God of battles. He who directs the storm, can preserve and defend." (APGTW p. 57) General Geary found himself in command of the Second Division, which was destined to become the famous White Star Division of the Twelfth Corps.

October 2, 1862. The Second Division of the 12th Corps at Loudoun Heights was presented for review that morning to President Lincoln who was accompanied by Generals Sumner and Franklin. It was accompanied by a 21 gun salute and all the honors. General Geary wrote to his wife, "Every thing went off satisfactory. Abraham looks quite care-worn and not nearly so well as he did when I last saw him. I am unable to say when we shall move from here, or to what position I shall be assigned. The Division consists of three Brigades but in consequence of the recent battle they are much reduced below their original strength." APGTW p. 59

October 8, 1862. The 102nd NY Volunteers were still at Loudoun Heights as part of the 12th Corps. There isn't much in the official reports about what was happening around Harpers Ferry at this time except by the letters written by General Geary and to some extent General Alpheus Williams who commanded the First Division and was first in line to command the Corps if the leading General was disabled, as sometimes happened. The official reports don't tell us much about the time between the Battle of Harpers Ferry on the 12 through 15th of September up until the end of the year. But General Geary wrote his wife: "The enemy are close in front of us and I capture some of his men daily. We have an immense army here, I cannot tell how many. That of the enemy is supposed to be 150,000 and Geary's Division still occupies the proud front of our lines. Some pretend to say the enemy contemplate a retreat. Others that they mean to fight. Be this as it may time alone can tell." APGTW p. 60

October 15, 1862. The 12th Corps is still at Loudoun Heights. Second Division Commander Geary had Knap's battery, which was partly commanded by his son Eddie, moved from Sandy Hook to their position at Loudoun Heights. Eddie was General Geary's older son, who at 16 years old became a second lieutenant in Knap's battery of artillery and in the summer of 1862 was promoted to first lieutenant and given a command of a two-gun section of his battery. He was an exceptionally attractive person and a highly competent officer. APGTW p. xviii

October 21, 1862. The Second Division under General Geary took 300 Calvary from the Sixth NY Calvary as well as Knap's Pennsylvania Battery, and the First and Second Brigades, leaving the third Brigade including the 102nd NY Volunteers behind, and made a reconnaissance "tour" to Lovettsville by way of Neersville to Hillsborough, then

Wheatland to Morrisonville, then finally to Lovettsville. Major General Burnside ordered the trip to Lovettsville. The only enemy found was a small group of White's Calvary Battalion under Captain R.B. Grubb.

They captured some rebel Calvary in a skirmish at the Glenmore farm one and a half miles north of Wheatland and pushed the rebels out of the area. General Geary reported, "Our Calvary exhibited much bravery in their charge, and throughout conducted themselves admirably. Colonel Devin, their Commander, Lieutenant-Colonel McVicar, and Major Carwardine, are deserving of much approbation for their display of gallantry and ability. In the charge Lieutenant-Colonel McVicar, who led it, had his horse shot under him, and received a bullet through his coat. The infantry and artillery evinced, upon the long march of 35 miles, a highly commendable spirit of alacrity, and great desire to be brought into action." WOTR v.19 part 2, pp 99-100. They camped at Lovettsville that night and returned to Loudoun Heights the next day.

October 26, 1862. The Union Army 12th Corps, with the 102nd NY Volunteers in the Second Division, was moved to Bolivar Heights, above Harper's Ferry. Bolivar Heights witnessed more battle action than any other location at Harpers Ferry. The first battle of the Civil War at Harpers Ferry occurred at Bolivar Heights on October 16, 1861, exactly two years after the commencement of the John Brown Raid. The formal surrender of nearly 12,700 Union soldiers, the largest capitulation of Union forces during the Civil War, occurred on Bolivar Heights later in the war. It was here the 12th Corps would stay for the winter of 1862-63, for the most part with a few reconnaissance missions in November. They were still recovering from the two big battles at Cedar Mountain and An-

tietam, and the loss of several of their leaders. At Cedar
Mountain, the 102nd lost Captains Julius Spring and Arthur
Cavanaugh and 21 enlisted men killed or mortally wounded,
Lieutenant Colonel Avery and six other officers and 70 enlisted
men wounded, and one officer and 14 men missing. At Antietam, it lost Captain Eugene Cornell and seven enlisted men
killed or mortally wounded, 24 men wounded, and five men
missing.

The largest encampment on Bolivar Heights occurred following the Battle of Antietam, in the fall of 1862. The entire
2nd Corps of the Army of the Potomac, numbering nearly
15,000 men, lived on Bolivar Heights for nearly six weeks before departing for the Fredericksburg Campaign during the
first week of November 1862. During this time, Professor
Thaddeus Lowe raised his gas-filled balloon to observe Confederate movements in the Shenandoah Valley. In addition,
the Union troops built extensive earthworks along the length
of the ridge, trying to make Bolivar Heights more defensible.
Bolivar Heights has a lower elevation than Maryland and
Loudoun Heights, which makes it easier to approach through
the Shenandoah Valley. Excerpts taken from: Bolivar Heights (U.S. National Park
Service). https://www.nps.gov/places/000/bolivar-heights.html

October 27, 1862. The 102nd NY Volunteers were transferred from the Second to the Third Brigade, and Gen. George
S. Greene assumed command of the Third Brigade. At age 61,
Greene was one of the oldest generals in the Union army and
his troops took to calling him "Old Man" or "Pap" Greene.
However, his age did not keep him from being one of the most
aggressive commanders in the army. He commanded the 3rd
Brigade, Second Division, II Corps, of the Army of Virginia at
the Battle of Cedar Mountain during the Northern Virginia

Bolivar Heights

Campaign. Attacked by a Confederate force three times the size of his own, Greene and his men refused to give ground, holding out until the neighboring Union units were forced to withdraw. His division commander, Brig. Gen. John W. Geary, received a severe wound during the action, and Greene temporarily took command of the division.

Greene was again temporarily elevated to command of his division, now designated part of the 12th Corps of the Army of the Potomac, at the Battle of Antietam. His division's three brigades were led by junior officers who had survived Cedar Mountain. Even though 12th Corps commander Brig. Gen. Joseph K. Mansfield was killed shortly after the fighting began, Greene led a crushing attack against the Confederates near Dunker Church, achieving the farthest penetration of Maj. Gen. Stonewall Jackson's lines of any Union unit. Under immense pressure, Greene held his small division – only 1,727 men engaged at the day's start– in advance of the rest of the army for four hours, but eventually withdrew after suffering heavy losses and low on ammunition. While the division was posted to Harpers Ferry, Greene took a three-week sick leave. Maj. Gen. Oliver O. Howard speculated that Greene, like many of his fellow officers, was sickened by the stench of dead and wounded at Antietam. When he returned, there was a new division commander, Brig. Gen. Geary. Greene was disgruntled that Geary, with only a few days seniority over him, was selected for the post; Geary had been wounded at Cedar Mountain, and his combat record was not as good, but his political connections and a sentiment that a wounded officer should not be unnecessarily set back in his career, gave him the appointment.

General Greene then resumed command of the 3rd Brigade, which included the 102nd NY Volunteers and Private Moses Whitbeck, and was involved in minor skirmishes in northern Virginia and not engaged at the Battle of Fredericksburg in December. Excerpts taken from https://db0nus869y26v.cloudfront.net/en/George_ S._Greene

October 28, 1862. The Second Division under General Geary moved the greater portion of the division to a valley situated immediately east of Loudoun Heights and Short Mountain. He would call it "Between the Hills," and wrote, "It is a beautiful place, easy of access and about the same distance from Harper's Ferry that the top of the mountain is. It is very warm and well sheltered from the storms by the surrounding mountains. You see I am still 'advance guard.'" APGTW p. 64

November 1, 1862. The 12th Corps, which included the 102nd NY Volunteers and Private Moses Whitbeck in the Third Brigade of the Second Division, remained at Bolivar

Heights while the Rest of the Union armies are preparing to advance toward Fredericksburg, Va. Bolivar Heights is a plateau that is 668 feet high, located near the towns of Bolivar, West Virginia, and the town of Harpers Ferry. During the American Civil War, it was the site of the October 1861 Battle of Bolivar Heights, during which Confederate States Army Colonel Turner Ashby failed to capture the heights from Union commander John White Geary. This is the same general who is in Command of the Second Division of the 12th Corps. In 1862, the Bolivar Heights were used by Confederate General Stonewall Jackson to shell Harpers Ferry, which was attacked during the Maryland Campaign. It was also attacked many other times, including Jubal Early's invasion of Maryland in 1864. Today, there are some cannons and a defensive trench preserved from the Civil War, and it is one of the designated historical locations of the National Park Service, part of the Harpers Ferry National Historical Park.

In 1862, that day was a Saturday, and General Geary would take a Balloon ride that day to scout enemy troops and last to get a feel of what this was like. General Geary wrote, "I made a Balloon ascension on Saturday in Professor Lowe's splendid balloon 'Intrepid,' for the purpose of ascertaining the precise location occupied by the enemy. I experienced no sensations of giddiness, and upon the whole if it had not been quite so windy, my aerial voyage would have been rather a pleasant one, as it was I could not complain of it. Having satisfied myself, and obtained the information I desired, I descended in safety to Terrafirma." APGTW p. 65

November 7, 1862. The 12th Corps was surprised by a snowstorm that dropped about 3 inches of snow. General Geary wrote his wife that changes were coming in military and

Cabinet circles. As it turns out, President Lincoln removed General McClellan from command of the Army of the Potomac this date and chose Major General Ambrose Burnside as the successor. Eddie Geary was now Quarter Master and Commissary for three Batteries under General Geary's command. APGTW p.67

November 9, 1862. A reconnaissance in force was made by the Union Army 12th Corps Second Division, which included Private Moses Whitbeck and the 102nd NY Volunteers, to Ripon, within 6 miles of Berryville, driving the enemy away and capturing prisoners, arms, horses and cattle, and ascertaining the location and strength of the enemy in the valley between there and Front Royal. About five miles from Rippon, General Geary discovered several Confederate camps that had been used by four or five regiments of the 12th Virginia Cavalry. The Union troops vigorously attacked the camps, forcing the Confederates to abandon the camps and retreat to Berryville. General Geary then returned to Harpers Ferry. General Geary wrote his wife, "God, in infinite mercy, has permitted me once more to return from an extended pursuit with the enemy, also another collision during a reconnaissance of 14 miles into Jefferson Co. Let us thank Him who giveth victory. For it is God only who can."

During the war, both Union and Confederate troops had marched into Berryville. Union units were stationed in the town from 1863 through 1864, including the 1st and 2nd Arkansas Cavalry, the 6th and 8th Missouri State Militia, and Gaddy's Home Guards. But the deadliest day happened on April 16, 1864, when Confederate guerrillas attacked a detachment of the 2nd Arkansas while they were foraging on the Osage branch of Kings River. Six Union soldiers and six black teamsters died during the fight. The Union casualties, and others who died at the post, are buried in the Berryville city cemetery. The Civil War was brutal to the town of Berryville. By 1865, most of the town lay in ruins.

November 26, 1862. The division made a second reconnaissance, with 600 infantry and 2 pieces of artillery, as far as Charlestown, having a skirmish with rebel Calvary at Cockrall's Mill, on the Shenandoah, routing them, wounding several, and taking a number of prisoners, arms, horses, together with a quantity of flour, and destroying at that place a cloth-mill that the rebels were using. They then marched to a point opposite Shannondale Spring and then to Charlestown, between which there and Halltown a rebel camp was broken up, and the 7th and 12th Virginia Calvary put to flight. No enemy was discovered, other than calvary parties, General Geary reported.

December 2, 1862. The Union Second Division, led by General Geary, reconned from Bolivar Heights toward Winchester with about 3200 infantry. They marched by Harpers Ferry and took the Winchester turnpike to Charlestown and arrived there about 8:30 a.m. They ran into two companies of the Twelfth Virginia Calvary, who had taken position in vacant homes and in woods about 0.75 miles mile outside of town. The Union forces routed them and pushed them out toward Charlestown, taking the Berryville Road, and as they pushed forward they pushed more calvary until they came upon the entire 12th Virginia Calvary, 200 or more strong. The rebel calvary charged the Second Division forces but were broken up by Knaps Battery and the 7th Ohio volunteers firing on them and dispersing them with demoralizing effect. Since it was now dusk, general Geary did not consider it sensible to advance any further as he wanted to let his men rest for the night and gain more information about rebel forces. The Division bivouacked there for the night and felled trees for temporary barricades.

December 3, 1862. The word had spread that A.P. Hill was in the area and that rebel forces were in nearby Millwood. Local residents confirmed that rebel troops were in Millwood the night before and also some at Newtown, South of Winchester. The bold actions taken by the rebel cavalry seemed to corroborate the information obtained. So General Geary took about 1000 infantry and four pieces of artillery and pushed toward Opequon Creek under occasional fire from rebel Calvary hiding in nearby woods. At Opequon, they discovered a camp that A.P. Hill's troops had vacated 3 or 4 days before. Here the advance part of the division waited for the rest of the division to come up and then they camped there for the night.

December 4, 1862. Early that morning, The 12th Corps Second Division, which included the 102nd NY Volunteers, cautiously moved toward Winchester, through Ash Hollow, keeping a heavy amount of flanking troops on each side of the dense woods they passed through. They forced rebel Calvary who were hiding in the thick woods to flee toward Winchester. As Geary's Division had gained the high ground that rose immediately east of the city, he ordered his troops into battle lines. At that time several people from the town or the other side came forward to inform General Geary that the rebel troops had evacuated the city the previous night. General Geary was unsure if the information was accurate, so he held them hostage and sent an attachment, including Captain R.C. Shannon from the 102nd NY Volunteers, who had risen to brigade assistant adjutant general, to the city to demand its surrender. Major Samuel B. Myers of the Seventh Virginia

Cavalry responded that they would within an hour's time depart and also make sure all the non-combatants who wanted to leave the town had that time to do so. General Geary rejected the one hour demand and started to circle the city to the north and east and to occupy forts on the north part of the city, when a flag of truce came. The mayor gave his unconditional surrender. Union forces then did a search of the city where they captured 118 rebel soldiers, a few Union soldiers who were prisoners, and a load of flour.

General Geary reported, "The illustration of the great revulsion of sentiment in favor of the Union was highly gratifying, and I beg to remark that our reception by the women and children was satisfactorily demonstrative. The outpouring of Union feeling was assisted with flags and other Union emblems. Most of the men being absent, a partial indication of the feeling prevailing was furnished by some 400 or 500 youths, whose acclimations of pleasure, beyond doubt unfeigned, were freely given. This change of feeling is similar through the country, and is strongly indicative of the growing entertainment of Union sympathies. Another subject worthy of comment is the destruction of Harper's Ferry and Winchester Railroad and of the property of the people, who have been bereft of nearly all the necessities of life. Devastation of a painful character is noticeable over all the section visited by the troops of Jackson and Hill. Cattle and hogs have nearly all been taken, and throughout a vast area there is not enough provender to maintain a troop of cavalry in any one neighborhood for a single week." WOTR Vol. 21 p. 34

Having stayed in Winchester until about 3 p.m., and considering their mission as completed, General Geary moved the division on the Martinsburg turnpike, and camped about six miles from Winchester. They did have some rebel Calvary shoot at them that night, but the confederates caused no damage and quickly left.

December 5, 1862. The Second Division marched and passed through Bunker Hill and Smithfield, and found the route in the same devastated condition as the land they had passed through the day before. At Oakland, about 75 calvary fired upon the Union outposts, but the men placed two pieces of artillery in a good spot in anticipation of the rebels and scattered them with a few well-placed shots.

General Geary reported, "The points of information gleaned upon the reconnaissance are summed up briefly, in effect that General D. H. Hill left with his division about November 17; Jackson with his command, about the 26th, and A.P. Hill from the 27th to the 29th of the same month. I am also led to believe, from various sources, that the combined forces of the enemy amounted to about 35,000 effective men and about 60 pieces of Artillery...No troops are now remaining in the valley, except Ashby's calvary. The forces driven from Winchester, when last heard from, were in full retreat beyond Strasburg. The two Hills and Jackson were last reported as marching directly toward Fredericksburg, and as within 20 miles of Lee's army." WOTR Vol. 21 p. 34

The Division bivouacked that night and the next (**December 6, 1862**) in the woods and endured a severe snowstorm without shelter but managed the short campaign without a single casualty. General Geary reported, "The expedition was prosecuted under circumstances the most disadvantageous, both as regards gathering information (which was an arduous, and in many instances impossible, case, except through immediate reconnoitering investigation) and the inclemency of the weather. A still further most noticeable drawback was in the scarcity of a calvary force, as I had but 50 of this arm, so essential to a vigorous persecution of the enemy, under the circumstances of their having so many mounted troops. Every movement required to be made slowly and with caution, supported by infantry. With a good regiment of Calvary, I might have captured the entire body of infantry and artillery opposing us, by cutting off their avenues of retreat in speedy maneuvering."

December 13, 1862. The 12th Corps Commander General H.W. Slocum notified Union Chief of Staff General Halleck that one of his division had arrived at Fairfax Courthouse and that the rest of the corps camped at Chantilly. General Slocum requested to remain there one day to have their artillery and calvary horses shod. The reply back was to move at once toward Fredericksburg without wagons and calvary. Because of this and the distance needed to travel to get there in time, with roads in bad condition, the 12th Corps would miss the battle of Fredericksburg from **December 13-15**. Colonel Van Buren was discharged on this date. He left the service due to poor health. He was brevetted Brigadier General by President Lincoln in 1865.

December 14, 1862. Lt. Colonel Lane was promoted to Colonel on this date. The 12th Corps left Chantilly that morning and struggling on poor roads they reached Fairfax Courthouse and halted there for a few hours. They then reached Fairfax Station around dark. The thawed snow and frost left the roads almost impassable. General Williams of the First Division wrote his daughter, "I am tonight trying to reduce the miles of wagons and reducing weight so as to pass through the sloughs ahead. You cannot imagine the difficulties of marches at this season, on short rations, short forage, bad roads, bad prep-arations, and the like. Think of moving a force of 10,000 men with all its supplies in wagons over a stripped country in the month of December… I am without intelligence from the active world for four days. We have rumors of great fighting at Fredericksburg and severe losses. I suppose we are bound for the same place, as we go now to Stafford Court House. I shall be there within four days." FTCM p. 153

December 15, 1862. The 12th Corps marched toward Wolf Run Ford over the Occoquan River, passed it, and continued a few miles past before camping at a crossroads by remnants of an old Baptist Church known as the Beacon Corner Church. It rained that night and into the next morning, saturating already wet roads.

December 16, 1862. The corps was up early and on the march. The roads were bad and almost impassable. Around noon the orders were given to stop and await further orders. Toward evening orders were given out for the corps to countermarch back to where they started. The roads they had used to get to where they made it, were really cut up and almost impossible to use. It would take the corps 2 more days to make it back to Fairfax camp. This would end operations for the remainder of the year as the roads were unusable. General Williams wrote his daughter: "Altogether I have never had so disagreeable and difficult a march. How my trains got over such roads I cannot now guess, but they have all returned safe, although shattered, with the loss of but one mule. I reached this place about 11 o'clock yesterday. Geary's Division is returning today. Where we are going is difficult to say. I suppose we shall remain hereabouts as a reserve until movements of the rebels are known." FTCM p. 154

1863

The Union Army 12th Corps was at Fairfax Station, Virginia, until **January 20, 1863**.

January 9, 1863 The Union Army 12th Corps was at Fairfax Station and the weather was clear and cold, and somewhat pleasant for winter. The two armies, The Union Army of the Potomac and the Confederate Army of Northern Virginia, were basically right across the Rappahannock River from each other. General Geary wrote his wife Mary, "Every thing is in the status quo here. Nothing new. The enemy are still before us. He may give us a fight at any time, if the dry weather continues, and if he does, we will give him the best shot in the locker." APGTW p. 81

January 12th, 1863 The Union Army 12th Corps with the Second Division was still in camp near Fairfax waiting for the winter to be over and the weather to get better. General Geary wrote, "The recent rains are making the roads exceedingly muddy and if it continues a few more days will render them impossible as the bottom soon falls out. From every thing that I can learn the enemy are contemplating a raid in great force north of the Rappahannock." APGTW p. 82

General Geary misinterpreted minor movements by the Confederates as each side was doing some minor skirmishing mostly done by Calvary.

January 16, 1863 The 12th Corps is still in Fairfax camp but are under orders to be ready to march at any notice with three days cooked rations on hand. General Geary wrote, "Every thing is ready, but nobody knows whither we go or for what purpose. The only thing is to be ready." APGTW p. 83

January 17, 1863 The weather was very cold, and many of the men of the Union Army 12th Corps had to bivouac, that is sleep without tents, because they were basically in a mobile situation and ready to make a movement at any time. General Geary wrote, "Tomorrow morning we leave this place under orders from gen. (sic) Burnside, and we will probably take the road once more to Dumfries, at which place I expect to pick up my first Brigade and a Battery of Artillery left there for the protection of that place. What our further destination will be, time alone can tell. The general opinion is we are again on the eve of stirring events." APGTW p. 83, 85

"Mud March" January 20-24, 1863. Following his defeat in the disastrous Battle of Fredericksburg in December 1862, Union General Burnside was desperate to restore his reputation and the morale of his Army of the Potomac. The day after Christmas, he began making preparations for a new offensive. This would involve feints at the fords upstream of Fredericksburg to distract the Confederates while he took the bulk of the Army across the Rappahannock River seven miles south of town. Finally, he planned for a cavalry operation on a grand scale, something that had never been done so far in the Eastern Theater, where so far Union horsemen had performed poorly and suffered repeated embarrassments at the hands of their Confederate foes.

The offensive began with a westward move on **January 20** in unseasonably mild weather. Burnside, with a head start, altered his plan and wanted to cross at Banks' Ford, a closer, quicker crossing. At dawn of **January 21**, engineers would push five bridges across; after that, two grand divisions would be over the river in four hours. Meanwhile, another grand division would distract the Confederate troops by repeating the December crossing at Fredericksburg. During the night of the **20th**, the rain began, and by the morning of the **21st**, the earth was soaked, and the riverbanks were a mess. Already, fifteen pontoons were on the river, nearly across it, and five more would complete the bridge.

Burnside started at once to bring up his artillery. For a significant area around the ford all day, the men worked in the rain but accomplished little. A large number of Union cannons were moved near the river, but **January 22nd** only added to the storm, and the artillery, caissons and even wagons were swamped in the mud.

The storm had hindered Burnside's movements, giving Confederate General Lee enough time to position the other side of the river with his army, though there was no attempt to interfere with his crossing except from the sharpshooters, who continued to fire on all occasions. No doubt Lee was hoping Burnside would complete a crossing, and with a swollen river in his rear, it would have been a sorry predicament for the Union Army. But Burnside finally became resigned to his fate and gave the order for the army to retire to its camps, and thus ended the famous mud march. Excerpts taken from Wikipedia: Mud March

On **January 23, 1863**, the Second Division went to Dumfries, but the 2nd and Third Brigade (which included the 102nd NY Volunteers and Private Moses Whitbeck) were five miles south at the Chopawamsic River.

The Mud March was Burnside's final attempt to command the Army of the Potomac. President Lincoln replaced him with Maj. Gen. Joseph Hooker on **January 26**.

After the "Mud March" came what some historians call "the strategic pause," which began on **January 25** and ended **April 27, 1863**. Union armies hit bottom in morale and fighting capacity following battlefield defeats in 1862 and faced a "landscape of defeat." The Army of the Potomac, "Mr. Lincoln's Army" of 135,000, settled down for a long, hard winter and built 30,000 huts; received better training and discipline; curbed desertion; and cared for its sick and wounded while its leaders fought off infighting, backbiting, political intrigues, congressional meddling, and popular discontent. They also confronted the Copperhead antiwar movement with unit resolutions, letters, and home front speeches. In camp, the army actively patrolled, gathered supplies, and made plans for continued operations and improved performance.

Some historians compare this time of the Union Army as their "Valley Forge." Fighting off demoralization, sickness, disease, and privation, the army's soldiers and animals persevered through the cruel and fickle weather. Every known historian suggests this was the army's low point in morale and effectiveness for the entire war in the East.

This strategic pause in the war allowed the army's leadership to restructure, reform, reorganize and develop plans for the coming spring offensive. During this time, they assumed a defensive posture, but conducted division-, brigade- and regimental-size raids, active reconnaissance, and logistical preparations. What resulted was a true non-battle turning point in the Union Army's wartime fortunes.

January 27, 1863 The Second Division was in Dumfries, Virginia. Dumfries was a small city situated on the Quantico River near its entrance to the Potomac. Located thirty miles from the Union capital of Washington, D.C., it was a strategic point that the Confederate Army hoped to control. The former port city of Dumfries, with its location roughly seventy miles from the Confederate capital of Richmond, was important to Union commanders. Prince William Forest Park (U.S. National Park Service). Civil War

General Geary wrote, "There is scarcely a probability that the army will move for several days, simply because it is impossible on account of the mud, which averages from 2 to 3 feet, or to the hub of a wagon. Oceans of it upon oceans. It has been snowing very rapidly all day, but it melts about as fast as it falls. Consequently, it has accumulated but little, and the mud is only increased." APGTW p. 87

January 30, 1863 The Union Army 12th Corps Second Division is still in Dumfries but preparing to march again. General Geary wrote his wife, "I drop you a hasty line from here, and if the weather is good I expect to advance tomorrow morning as far as Stafford Ct House. The roads are near impossible as it is possible to conceive of anything to be, but nevertheless I will try them, and if I try, my motto is success... The snow and mud are about 16 inches deep on average. Consequently the whole earth is on the move." APGTW p. 87

February 2, 1863 The Second Division is camped at Stafford Courthouse, about 7 miles from Aquia landing where their army drew all of its supplies. The majority of the army is camped at Falmouth, about 10 or 11 miles away. In his last letter to his wife until the end of May General Geary wrote, "The health of the troops is generally good, considering the inclemency of the weather. The appointment of General Hooker is more popular that I at first apprehended it would be, and should he succeed he will knock the wind out of some peoples sails that I could name. This is a possibility, for Hooker will fight "to whip" when he makes a start, and you may look out either for a great victory, or for a corresponding defeat when he strikes." APGTW p. 89

February 5, 1863 General Joseph Hooker reorganized the Army of the Potomac. Following Fredericksburg, General

Burnside was removed from command of the army and replaced by Joe Hooker. Hooker immediately abolished the Grand Divisions and also for the first time organized the cavalry into a proper corps led by George Stoneman instead of having them unsuccessfully distributed among infantry divisions. Burnside and his old IX Corps departed out to a command in the Western Theater. The I, II, and 12th Corps retained the same commanders they had had during the Fredericksburg campaign, but the other corps got new commanders once again. Daniel Butterfield was chosen by Hooker as his new chief of staff, and command of the V Corps went to George Meade. Daniel Sickles received command of the III Corps and Oliver Howard the XI Corps after Franz Sigel had resigned, refusing to serve under Hooker, his junior in rank.
Seeing the Civil War -Army of the Potomac

February 16, 1863 Although General Geary ignored the time of February until after the battle of Chancellorsville in May in his letters to his wife, General Williams who commanded the First Division of the 12th Corps did not. General Williams wrote to his daughter about Stafford Courthouse, "Just over the hills east is what is called Stafford Courthouse. One old tavern, awfully used, one small courthouse, windowless, one smaller jail, a blacksmith shop for mule shoeing, and one tolerably looking house terraced in front, where 'Mit Sigel' has his headquarters and around which barbaric Dutch is uttered with most villainous vehemence and gutturalness." FTCM p. 165

February 22, 1863 The weather in Stafford Courthouse was stormy with snow and high winds. The 12th Corps had hundreds of men on picket duty and thousands poorly sheltered. Firewood was being consumed at a large rate but the supply was hard to get. The men had a large celebration, anyway, as it was Washington's birthday.

March 7, 1863 The 12th Corps is still near Stafford Courthouse and in preparation for the spring campaign is having reviews by General Hooker and the generals in his command, weather permitting.

General Williams wrote, "I have 2,000 men on duty daily on pickets, road building, guards, etc., and the mud is so deep yet that we can hardly fix up. For a season I am obliged to cram my whole division into a space about big enough for a regiment, such is the broken nature of the country, full of deep ravines and rolling hills." FTCM p. 167

March 9, 1863 A Confederate raid led by Captain John S. Mosby (with only 29 men) swoops in on Fairfax Courthouse, Virginia, capturing Union General Edwin Stoughton in his nightshirt. In addition to General Stoughton, there are two other officers and 30 enlisted men captured as well as 58 horses. There are no Confederate casualties. Supposedly when told of the raid, President Lincoln said that he could always create another brigadier general, ."..but those horses cost $125 apiece!"

Mosby wrote in his memoirs that he found Stoughton in bed and roused him with a "spank on his bare back." Upon being so rudely awakened, the general indignantly asked what this meant. Mosby quickly asked if he had ever heard of "Mosby." The general replied, "Yes, have you caught him?" "I am Mosby," the Confederate ranger said. "Stuart's cavalry has possession of the Courthouse; be quick and dress." Mosby was formally promoted to the rank of captain two days later, and major on March 26, 1863. Mosby's Rangers are on their

way to becoming a legend and Mosby himself becomes the
"Grey Ghost." Infogalactic: John S. Mosby

March 10, 1863 President Lincoln offers amnesty to
thousands of Union deserters if they will surrender and return
to their units. By this time, the armies of both sides have massive desertion rates, as the war drags on.

Lincoln proclaimed, "In pursuance of the twenty-sixth
section of the act of Congress, entitled an act for enrolling and
calling out the National forces, and for other purposes, approved on the third of March in the year one thousand eight
hundred and sixty-three, I ABRAHAM LINCOLN, President
and Commander-in-Chief of the Army and Navy of the United
States do hereby order and command that all soldiers enlisted
or drafted into the service of the United States, now absent
from their regiments without leave, shall forthwith return to
their respective regiments and I do hereby declare and
proclaim that all soldiers now absent from their respective
regiments without leave who shall on or before the 1st day of
April, 1863, report themselves at any rendezvous designated by
the General Orders of the War Department, No. 58, hereto
annexed, may be restored to their respective regiments without
punishment except the forfeiture of pay and allowances during
their absence, and all who do not return within the time above
specified, shall be arrested as deserters, and punished as the law
provides:

"And whereas, evil disposed and disloyal persons, at sundry
places, have enticed and procured soldiers to desert and absent
themselves from their regiments, thereby weakening the
strength of the armies and prolonging the war, giving aid and
comfort to the enemy, and cruelly exposing the gallant and

faithful soldiers remaining in the ranks to increased hardships and dangers;

"I do, therefore, call upon all patriotic and faithful citizens to oppose and resist the aforementioned dangerous and treasonable crimes, and aid in restoring to their regiments all soldiers absent without leave, and to assist in the execution of the act of Congress for "enrolling and calling out the national forces and for other purposes," and to support the proper authorities in the prosecution and punishment of offenders against said act and aid in suppressing the insurrection and the rebellion." The American Presidency Project - Recalling Soldiers to Their Regiments

March 10-April 4, 1863 The 102nd NY Volunteer Regiment is detached in New York on special duty. There is very little information about this (General Geary's book skips March and April and doesn't pick back up until the end of May), but maybe as part of the strategic pause the regiment went back to their home state for some R&R and also to rally the people with "homefront" speeches to help the Union cause. Also, the New York State Record shows that the 102nd NY were part of an allowance made to the State of New York to have some of their regiments serve the state for periods of time no more than 100 days. New York in the War of the Rebellion, p. 58

What was the 102nd NY doing in the spring of these war years 1863-1865? Sometimes records are incomplete. There are many sources to comb through; some are official records and some information is in letters written by soldiers that were kept by their families and made into books. There are also countless diaries in private collections that are not available to the public. I'm hoping that by publishing this book someone will come forward with more information about the 102nd NY Volunteers. Most of the information found about early 1863 indicate that the Regiment was in New York from **March 10** until **April 4**. Unfortunately, the War of the Rebellion Series is mostly reports and correspondences of major engagements during battles or on the way there or afterwards, and it is sometimes hard to find information about less active times. Also, General Geary's book omits any letters between early February until late May, entirely skipping the battle of Chancellorsville and the time leading up to it. Also, at times army units were rotated out of battle ready status so they could recover from a battle or be reorganized if they lost key leaders. This also afforded the soldiers some R&R time and short furloughs. Sometimes these units were assigned duty in the rear areas where they would guard key military assets while they recovered or were reorganized.

Civil War winters were particularly trying and monotonous for the armies. Impassable, muddy roads and harsh weather precluded active operations. Disease ran rampant, killing more men than battles. But with all of its hardships, winter also allowed soldiers an opportunity to bond, have a bit of fun, and enjoy their more permanent camps. Through these dreary months all soldiers, Union and Confederate, had to keep warm and busy to survive.

While on the move in warmer weather, soldiers often slept in easily-erected canvas tents or they simply slept without cover under the stars. In the winter, large camps were established with more substantial shelter. Winter huts were built by the armies out of the surrounding materials including trees, mud, leaves, and soldiers' canvases. These huts usually included a chimney, which kept the small space warm, but some were more effectively built than others.

The camps were set up much like small villages complete

with crisscrossing lanes called "company streets," churches, and sutlers' shops. While this may seem cozy, these temporary villages lacked the appropriate systems to provide clean water and clear away waste; in addition, food was scarce. Disease and death abounded and spread easily.

March 20, 1863 The 12th Corps is still at Stafford Courthouse, and the weather is still cold and bad with snow falling, the men are still preparing for the spring campaign with reviews and inspections frequently. General Williams of the first Division wrote, "Yesterday General Hooker reviewed my division and pronounced it a 'splendid division.' The regiments certainly never looked better or did better. The day before I was reviewed by Gen. Slocum, and the day before that I reviewed a brigade. Inspections of regiments go on when it don't rain or snow. There are so many things to look after by the way of supplies before taking the field that everybody must be active, each one inspecting to see that those below are doing their duty." FTCM p. 169

Chancellorsville Campaign April 27-May 6, 1863

At the Battle of Chancellorsville in May 1863, Greene's brigade, which included the 102nd NY Volunteers, was in the center of the line. When the Union right—the XI Corps—collapsed, Greene's Brigade was subjected to enfilade artillery fire and then infantry assaults. General Greene had ordered his men to fortify their positions 200 yards to their front using abatis and trenches and they were able to hold out against several Confederate assaults, although losing 528 men of 2,032 engaged. During part of the battle, Greene once again assumed temporary command of the entire division when General Geary was wounded again.

April 26, 1863 The 102nd NY Volunteers Regiment left Aquia Creek, Virginia, and marched 17 miles and stopped for the night. Aquia Creek was part of Stafford County where the 12th Corps had spent the winter. It is located north of Fredericksburg on a bay west of the Potomac River that feeds into the river and was an important place for commerce during the war.

Frustrated with the horrible defeat inflicted on his army by Lee's jubilant Confederates, President Lincoln replaced General Burnside with General "Fighting Joe" Hooker. Rumors were that General Lee was planning an immediate attack on the Union Army, and as Hooker took command and reviewed the position of his troops, he saw that there was a major gap in his lines and that his key supply depot at Aquia Creek was in danger.

The loss of the supply base at Aquia Landing would be devastating to the Union Army and easily could lead to its defeat and withdrawal from Northern Virginia.

To counter this danger, Gen. Hooker ordered the XI Corps of Gen. Oliver O. Howard to move its camps to a new position along the heights overlooking Accokeek Creek. This would "plug the gap" in the Union lines and protect Aquia Landing.

April 27, 1863 The 12th Corps Second Division marched toward Stafford Courthouse. They passed through Summer Duck, or Crittenden's Mills. They stopped for the night at a farm some three miles east of Hartwood Church having marched 14 miles. The Hartwood Church was organized in June 1825 by the Winchester Presbytery as Yellow Chapel Church, the brick church was constructed between 1857 and 1859. It became Hartwood Presbyterian Church in 1868. During the Civil War an engagement took place there on

February 25, 1863. Confederate Brig. Gen. Fitzhugh Lee, commanding detachments of the 1st, 2d, and 3d Virginia Cavalry Regiments, defeated a Union force and captured 150 men. The interior wooden elements and furnishings of the church suffered considerable damage during the war but were replaced. The building was listed on the Virginia Landmarks Register and the National Register of Historic Places in 1989, and it is an American Presbyterian Reformed Historical Site.
Historical Marker Project. Hartwood Presbyterian Church - Fredericksburg - VA - US

April 28, 1863 The Second Division marched toward Kelly's Ford on the Rappahannock River. Kelly's Ford had been the site for a battle in **March 1863**. At the time, the recently organized Union Cavalry Corps possessed superior equipment and the advantages of a plentiful supply of men and horses over their Confederate counterparts, but they lacked the confidence, experience, and leadership to challenge Maj. Gen. J. E. B. Stuart's troopers. That March, Confederate cavalry under Brig. Gen. Fitzhugh Lee conducted raids against elements of the Union line along the Rappahannock River. In response, on **March 16th**, 2,100 Union cavalrymen under Brig. Gen. William Averell set out to "rout or destroy" Lee and his cavalry as the Confederates sheltered south of the river near Culpeper Courthouse. The next day, Averell forced a crossing at Kelly's Ford, 25 miles upstream from Fredericksburg, and pressed forward two miles into open ground. Averell repulsed several of Lee's attacks, forcing the Confederate cavalry from the field with a counterattack. Stuart's youthful and "gallant" artillery chief, Major John Pelham, was killed. Despite having victory in his grasp, Averell withdrew back across the river that evening. Lee had avoided being crushed, which meant that Averell had failed in his primary objective, but the battle proved that Union cavalry was a force to be reckoned with. The battle set the stage for Brandy Station and other cavalry actions of the Gettysburg Campaign that summer. American Battlefield Trust. Kelly's Ford Battle Facts and Summary

April 29, 1863. The Union Army 12th Corps under the command of General Henry W. Slocum, with Division Commanders Brigadier General Alpheus S. Williams and Brigadier General John W. Geary was on the march toward Chancellorsville, Virginia. The 12th Corps was 13,450 men strong. The 102nd NY Volunteers with Private Moses Whitbeck were in Geary's division, but General Geary's letters ignored this particular battle. General Williams wrote a detailed account of the actions taken by the 12th Corps as he became their temporary commander for 3 days. The 12th Corps were the first to cross the Rappahannock River and were nearing the Rapidan River at Germanna Ford when they heard the sound of cannons and small arms fire to the south. The whole corps was massed in columns on the low ground bordering the river, around which there was an amphitheater of hills surrounding it. Across the river, long lines of infantry were winding down toward the river, and on the south side brigades were breaking ranks and deploying on the hills. Batteries of artillery and large columns of calvary were forming large groups. The 12th Corps crossed the river, but not until after midnight did all the men get across. The men were wet but in good spirits. The corps at this time contained 30 regiments of infantry, with five batteries of light artillery, numbering in all 19,929 available for duty. The Battle of Chancellorsville was upon them. – The Civil War Months: Hooker Moves in Earnest

April 30, 1863 It was a dismal, drizzling morning as the Union Army of the Potomac, commanded by Maj. Gen.

Joseph Hooker which included the 12th Corps, crossed the Rappahannock and Rapidan rivers as planned and began to concentrate around the hamlet of Chancellorsville, which was little more than a single large, brick mansion at the junction of the Orange Turnpike and Orange Plank Road. Built in the early 19th century, it had been used as an inn on the turnpike for many years but now served as a home for the Frances Chancellor family, thus the name Chancellorsville. Academic.com: Battle of Chancellorsville.

Some of the family remained in the house during the battle. The 12th Corps, including the 102nd NY volunteers and Private Moses Whitbeck, were advancing carefully toward Chancellorsville led by General Geary's Division. The road was an old worn out plank road full of gullies and holes and slippery from the rain. The march was hard, but the men took it well, and there was no straggling. The Corps reached Chancellorsville around 3 p.m. Colonel Lane reported, "Marched from 7:30 a.m. until 4pm toward Fredericksburg, Third Brigade in advance, with the One hundred and second New York Volunteers on the left, when we were filed to the right, formed in line of battle, and, with skirmishers thrown out, advanced through the trees about one-eight of a mile to a narrow wagon road, when we halted for the night, the men lying on their arms." WOTR Vol 25, p. 764

The Chancellorsville campaign was one of the most lop-sided clashes of the war, with the Union's effective fighting force more than twice the Confederates, the greatest imbalance during the war in Virginia. Hooker's army was much better supplied and was well-rested after several months of inactivity. Lee's forces, on the other hand, were poorly provisioned and scattered all over the state of Virginia. Some 15,000 men of

Confederate General Longstreet's Corps had previously been detached and stationed near Norfolk in order to block a potential threat to Richmond from Federal troops stationed at Fort Monroe and Newport News on the Peninsula, as well as at Norfolk and Suffolk.

In light of the continued Federal inactivity, by late March Longstreet's primary assignment became that of requisitioning provisions for Lee's forces from the farmers and planters of North Carolina and Virginia. As a result of this the two divisions of Maj. Gen. John Bell Hood and Maj. Gen. George Pickett were 130 miles away from Lee's army and would take a week or more of marching to reach it in an emergency. After nearly a year of campaigning, allowing these troops to slip away from his immediate control was Lee's gravest miscalculation. Although he hoped to be able to call on them, these men would not arrive in time to aid his outnumbered forces. Cloudfront.net: Battle of Chancellorsville

As far as Union General Slocum's 12th Corps was concerned, they formed in a half circle of sorts to the south and west near a plateau named Hazel Grove. Geary's Division was on the Union left (East) and Williams' division was on the right. General Williams wrote, "It was a pleasant moonlight night. Chan-cellorsville house became the center of hundreds of officers. It was a gay and cheerful scene…everyone prophesied a great success, an overwhelming victory…All was couleur de rose! How many joyous hearts and bright cheerful faces beat and smiled happily for the last time on that delightful moonlight night as Chancellorsville!" FTCM p. 185

Battle of Chancellorsville May 1-5, 1863
May 1, 1863 The morning started foggy but soon dawned

clear and cool as Union General Hooker's Army of the Potomac troops rose from their night's sleep and assembled themselves around their morning campfires. The aroma of coffee permeated the chilly morning air as they awaited the command to march. Hooker's corps commanders were becoming increasingly impatient. They wanted to move out of the wilderness area that surrounded Chancellorsville and onto more advantageous ground to meet Lee's ever-menacing army. Historynet.com: Battle of Chancellorsville.

General Williams wrote, "I never saw my troops in better condition, more anxious to meet the enemy." FTCM p. 186 His troops had been marching an average of 15 miles per day over hard, muddy roads and carrying about 60 pounds on their backs but still where not discouraged and eager to take on the rebels.

Around 11 a.m., Hooker finally passed the word to proceed along several routes eastward. The 12th Corps would mass below the plank road and advance in small parties to conceal themselves from the enemy, toward Tabernacle Church. XI Corps would follow about a mile to the rear of 12th Corps; one division from II Corps would take up positions near Todd's Tavern; III Corps would consolidate on the United States Ford Road about a mile from Chancellorsville; and Brig. Gen. Alfred Pleasanton would keep his cavalry detachments at Chancellorsville.

As Hooker advanced from Chancellorsville toward Lee, the Confederate general split his army in the face of superior numbers, leaving a small force at Fredericksburg to deter Maj. Gen. John Sedgwick from advancing, while he attacked Hooker's advance with about four-fifths of his army. The two forces met near the Zoan Church, three miles east of Chancellorsville, late that morning. On the Orange turnpike, the Union Fifth Corps encountered Confederate Maj. Gen. Lafayette McLaws' division and after three hours of fighting was pushed back. Elements of the Twelfth Corps, including the 102nd NY Volunteers, were likewise slowed by Confederate Lt. Gen. Richard Anderson's Division on the plank road to the south, but 10 regiments were in the woods close to the road, and the men were eager and cheerful to engage the Confederates.

Then, inexplicably, Hooker ordered his corps commanders to fall back to Chancellorsville, despite the objections of his subordinates. Hooker withdrew his men to the defensive lines around Chancellorsville, ceding the initiative to Lee. Historians debate why Hooker pulled back when he had the advantage. Some think he lost his nerve in the face of General Lee, who was an imposing commander. Others believe his actions may have demonstrated his lack of confidence in handling the complex actions of such a large organization for the first time (though he had been an effective and aggressive division and corps commander in previous battles), but he had also decided before beginning the campaign that he would fight the battle defensively, forcing Lee, with his small army, to attack his own, larger one.

At the Battle of Fredericksburg on **December 13, 1862**, the Union Army had done the attacking and met with a bloody defeat. Hooker believed it was better to have Lee to attack him, and Lee would oblige him. As darkness approached, both armies dug in for the night. Morale in the Army of the Potomac was sinking rapidly. One soldier in Meade's corps remarked, "All enthusiasm vanished, all the bright hopes of success disappeared." On the Confederate side, Lee and Jackson met on the Plank-Furnace crossroads to conceive a battle plan for the next day. Excerpts taken from History Maps.com: Battle of Chancellorsville

As for the 102nd NY Volunteers, Colonel Lane reported, "At 1 o'clock fell in, marched by the left flank to the plank road, and, together with the rest of the Twelfth Corps, made a reconnaissance 2 miles to the front, when the corps was formed in line of battle, the Third Brigade on the right of the plank road. Skirmishers were thrown out, and the brigade advanced half a mile without meeting the enemy, when we found ourselves under a heavy fire from the rebel batteries. The line was here halted, and, after remaining about 15 minutes, were faced to the rear and marched back to the ground of the first line of battle. Here the command rested on their arms for nearly an hour, when we were again marched to our camping ground of the previous night; cooked supper, and after dark formed rifle pits of logs, with abatis in front, and filled in outside with dirt from trenches. These trenches were dug by the bayonets of the men, and the dirt removed by their tin cups and plates. Six miles of rifle pits were reported to be completed in this army at sunrise next morning, and mostly completed without intrenchment tools." WOTH Vol 25 p. 764

As for the 102nd NY Volunteers in the Third Brigade, General Geary reported, "The conduct of Greene's Brigade was admirable at this juncture. Although exposed for quite a length of time to the fire of the enemy in a position where they could neither shelter nor defend themselves, nor return the assault, they bore themselves with the calmness and discipline of veterans, emulating the example so ably given them by their brigade commander. The fire of the enemy slackened after some half hour's play upon our line, and I then received orders to fall back to my original position near Chancellorsville. This was accomplished in good order by the whole command, notwithstanding that a harassing attack was continued by the enemy upon our left flank almost up to the line of our defenses." WOTH Vol 25 p. 729

On **May 2, 1863**, the morning began with dense fog, and the troops had their breakfast before sunrise and the small amounts of tents they had used the night before were packed up. The 12th Corps had built a breastwork of logs in their front.

Lee divided his army again, sending Stonewall Jackson's entire corps of about 30,000 men on a flanking march that clandestinely crossed the front of the Union Army and swung around behind it. Jackson's objective was the right flank of the Union line that rested "in the air" along the Orange Turnpike near Wilderness Tavern. At about 5 p.m., Jackson, having completed his circuit around the enemy, unleashed his men in a violent attack on Hooker's right and rear. While members of the XI Corps in the Army of the Potomac were settling down to fix supper, Jackson's men burst out of the thickets screaming the rebel yell. They shattered the Federal Eleventh Corps and pushed the Northern army back more than two miles. Excerpts taken from American Battlefield Trust: Chancellorsville Articles

The Union Army really should not have been surprised as some scouts had climbed a tall tree and seen the columns of rebels passing across the Orange Courthouse road and massing to the right of the 11th Corps. Geary's Division, including the 102nd NY Volunteers, had been sent out on the plank road earlier in the day but had met stiff resistance and fell back to their entrenchments. When the 11th Corps was attacked, the men of the 12th Corps were there as the fleeing 11th Corps was pushed into them. General Williams wrote, "I saw at once that all effort to organize such a body of men was fruitless.

They were like a flock of scared sheep driven into a corner; not one thought of defense. The crack of the musket was close at hand." FTCM p. 191

Two of his brigades of the 1st Division of the 12th Corps came to the rescue and checked the advance of Confederate General Jackson's force. But then surprisingly General Hooker ordered the 12th Corps to fall back to their previous entrenchments. General Williams saw this as a danger and wrote, "No one could tell friend from foe nor see a hidden enemy a rod away." Much confusion was caused, the fighting continued, and many died because of it. General Williams wrote, "Human language can give no idea of such a scene; such an infernal and yet sublime combination of sound and flame and smoke, and dreadful yells of rage, of pain, of triumph, or of defiance." FTCM p. 192

The battle raged in the dark for several more hours until suddenly it all stopped, as no one could see. Each side lay quiet listening to determine where the other side was. The fighting was over for the night, and the midnight air became unpleasantly cold.

While performing a personal reconnaissance in advance of his line that evening, Confederate General Jackson was wounded by fire after dark from his own men close between the lines during all this confusion. A North Carolina regiment opened fire, mistaking them for enemy cavalry. A bullet struck Jackson, shattering the bone above his left shoulder. Cavalry commander Maj. Gen. J. E. B. Stuart temporarily replaced him as corps commander. Jackson would die several days later, following amputation of his left arm.

Colonel Lane reported, "Remained in rifle-pits, with occasional picket skirmishes with the enemy. In the afternoon witnessed the breaking of the Eleventh Corps to our right and rear. The battery to the left of the Eleventh Corps, which was on a prolongation of our line to the right, had been firing heavily at the rebels marching past the front of the Twelfth Corps during the earlier part of the day, and apparently occupied its own ground at dark." WOTR v. 25 p. 764

May 3, 1863 Despite the fame of Confederate General Stonewall Jackson's victory on **May 2**, it did not result in a significant military advantage for the Army of Northern Virginia. The battle of Chancellorsville continued, as the long marches and daring tactics of the last two days gave way to a slugging match in the impenetrable woods on three sides of the Chancellorsville intersection. The fighting was intense, and the casualties mounted on both sides. Union General Howard's XI Corps had been defeated, but the Army of the Potomac remained a potent force, and Union General Reynolds's I Corps had arrived overnight, which replaced Howard's losses. About 76,000 Union men faced 43,000 Confederates at the Chancellorsville front. The two halves of Lee's army at Chancellorsville were separated by Union General Sickles' III Corps, which occupied a strong position on high ground at Hazel Grove. Unless Lee could devise a plan to remove Sickles from Hazel Grove and combine the two halves of his army, he would have little chance of success in assaulting the formidable Union earthworks around Chancellorsville. Fortunately for Lee, Union General Joseph Hooker inadvertently cooperated. Early on **May 3**, Hooker ordered Sickles to move from Hazel Grove to a new position on the plank road. As they were withdrawing, the trailing elements of Sickle's corps were attacked by the Confederate brigade of Brig. Gen. James J. Archer, which captured about 100 prioners and four cannons. Hazel Grove was soon turned into a powerful artillery platform with 30 guns under Col. Porter Alexander. Confederate artillery roared from Hazel Grove, and Southern infantry persistently pushed ahead. When a Confederate artillery round smashed into a pillar against which Hooker was leaning, the Union leader was knocked unconscious for a half hour. His return to semi-consciousness disappointed the veteran corps commanders, who had hoped that without him they would be free to employ their army's considerable untapped might.

Although Stuart, who replaced a wounded Jackson, was a cavalryman who had never commanded infantry before, he delivered a creditable performance at Chancellorsville. By the morning of **May 3**, the Union line resembled a horseshoe. The center was held by the III, XII (including the 102nd NY Volunteers and Private Moses Whitbeck), and II Corps. On the left were the remnants of the XI Corps, and the right was held by the V and I Corps. On the western side of the Chancellorsville salient, Stuart organized his three divisions to straddle the plank road: Heth's in the advance, Colston's 300–500 yards behind, and Rodes', whose men had done the hardest fighting on **May 2**, near the Wilderness Church. Excerpts taken from History-maps.com: Battle of Chancellorsville

As for the 102nd NY Volunteers, Colonel Lane reported, "At about 10 a.m. the brigade, being under a heavy enfilading fire from the enemy's batteries, was ordered to fall back in good order. The enemy were in large numbers, coming toward us through the trench, under cover, but below the line of fire of their artillery, and were strongly pressing the most advanced regiments with a galling fire of musketry, throwing them somewhat in disorder. The One hundred and Second being within rifle shot, left the trenches and formed at right angles to it, and poured volleys of musketry into the advancing rebels, which halted them, giving these regiments time to withdraw. After they had passed, rebels came in on both sides, left and right, saying we were surrounded and must surrender, but instead of doing so we disarmed 2 commissioned officers, 1 flag sergeant, and 20 privates, taking the flag, and bringing our prisoners safe to the rear. The battle flag and prisoners were from the Twelfth Georgia Volunteers." WOTR Vol 25 p. 765

By midday after four hours of heavy fighting, Southern infantry smashed through the final resistance and united in the Chancellorsville clearing. Their boisterous, well-earned, celebration did not last long. Word came from the direction of Fredericksburg that the Northern rearguard was threatening the Confederate army's rear. Lee had left a relatively small force at Fredericksburg, ordering Brig. Gen. Jubal Early to "watch the enemy and try to hold him." If he was attacked in "overwhelming numbers," Early was to retreat to Richmond, but if Sedgwick withdrew from his front, he was to join with Lee at Chancellorsville.

By midmorning, two Union attacks against the infamous stone wall on Marye's Heights were repulsed with numerous casualties. A Union party under flag of truce was allowed to approach ostensibly to collect the wounded, but while close to the stone wall, they were able to observe how sparsely the Confederate line was manned. A third Union attack was successful in overrunning the Confederate position. Early was able to organize an effective fighting retreat, and Union General Sedgwick's road to Chancellorsville was open, but he wasted time in gathering his troops and forming a marching column. His men, led by Brooks' Division, followed by

Newton and Howe, were delayed for several hours by successive actions against the Alabama brigade of Brig. Gen. Cadmus M. Wilcox. His final delaying line was a ridge at Salem church, where he was joined by three brigades from McLaws' Division and one from Anderson's, bringing the total Confederate strength to about 10,000 men.

Artillery fire was exchanged by both sides in the afternoon and at 5:30 p.m., two brigades of Brooks' Division attacked on both sides of the plank road. The advance south of the road reached as far as the church yard but was driven back. The attack north of the road could not break the Confederate line. Wilcox described the action as "a bloody repulse to the enemy, rendering entirely useless to him his little success of the morning at Fredericksburg." Hooker expressed his disappointment in Sedgwick: "My object in ordering General Sedgwick forward ... Was to relieve me from the position in which I found myself at Chancellorsville. ... In my judgment General Sedgwick did not obey the spirit of my order and made no sufficient effort to obey it. ... When he did move it was not with sufficient confidence or ability on his part to manoeuvre his troops." American Battlefield Trust: Chancellorsville Articles

As for the 102nd NY Volunteers, Colonel Lane reported, "Having become detached from the brigade, we formed line (at the order of an officer of General Geary's staff) on the left, and in front of the battery at the burning brick house, in order to pour in a flanking fire upon the rebels should they attempt advancing through the meadow to take the battery. We were driven from this point by the fire of our own artillery, and then at the insistence of Lieutenant-Colonel Dickinson, of General Hooker's staff, we took position directly behind the battery and supported it until it retired and remained in the same position until another battery replaced the retiring one, which, after firing some rounds of shell and round shot, also retired. After the battery moved, we were the extreme left at that point, as far as we could see, except quite a number of stragglers, who had formed on our right and left, when we again marched under artillery fire to rejoin our brigade and reached it soon after. In a few minutes thereafter, the brigade joined the division, halted and rested for 10 minutes. The rebels shelled us, probably seeing the smoke from the fires made by our men for cooking purposes. The command was moved farther to the right, and then countermarched by the right flank and marched to the road, and soon after took position in rifle-pits about 2 miles from the United States Ford. He we encamped for the night, the One hundred and second New York Volunteers acting as a reserve." WOTR Vol 25 Pg 765

On the evening of **May 3**, Hooker remained in his defenses north of Chancellorsville. Lee observed that Hooker was threatening no offensive action so felt comfortable ordering Anderson's Division to join the battle against Sedgwick. He sent orders to Early and McLaws to cooperate in a joint attack, but the orders reached his subordinates after dark, so the attack was planned for May 4.

The fighting on **May 3, 1863**, was some of the most furious anywhere in the Civil War. The loss of 21,357 men that day in the three battles, divided equally between the two armies, ranks the fighting only behind the Battle of Antietam as the bloodiest day of war in American history.

On **May 4, 1863**, the Union Army 12th Corps with the 102nd NY Volunteers and Private Moses Whitbeck were on some bluffs above Mineral Spring (now called the Pike Dam Run) with the Rappahannock River to their left (East). The men were busy making rifle pits and were determined to make an impregnable position to keep their retreat options open to the U.S. Ford on the Rappahannock.

Colonel Lane reported, "At 1 o'clock, May 4, we were marched to the left and rear, and occupied the heights commanding the ford. The One hundred and second then made rifle-pits and trenches the whole length of the regiment, with logs, abatis, &c." WOTR Vol 25 p. 765

Meanwhile over near Fredericksburg, just a few miles away, Confederate General Early's plan was to drive the Union troops off Marye's Heights and the other high ground west of Fredericksburg. By this time Union General Sedgwick had placed his divisions into a strong defensive position with its flanks anchored on the Rappahannock, three sides of a rectangle extending south of the plank road. Lee ordered McLaws to engage from the west "to prevent [the enemy] concentrating on General Early." Early reoccupied Marye's Heights in the morning, cutting Sedgwick off from the town. However, McLaws was reluctant to take any action. Before noon, Lee arrived with Anderson's Division, giving him a total of 21,000 men, slightly outnumbering Sedgwick. Despite Lee's presence, McLaws continued his passive role, and Anderson's men took a few hours to get into position, a situation that frustrated and angered both Early and Lee, who had been planning on a concentrated assault from three directions.

The attack finally began around 6 p.m. Two of Early's brigades (under Brig. Gens. Harry T. Hays and Robert F. Hoke) pushed back Sedgwick's left-center across the plank road, but Anderson's effort was a slight one and McLaws once again contributed nothing. Throughout the day on May 4, Hooker provided no assistance or useful guidance to Sedgwick, and Sedgwick thought about little else than protecting his line of retreat. McLaws and Early's counterattack pushed Sedgwick back across the river, halting the Union threat from the east. Academic.com: Battle of Chancellorsville.

As for the 12th Corps, the attack from the rebels never came. General Williams wrote: "Nothing was seen of the enemy except a few cavalry and infantry pickets, with which ours exchanged occasional shots." FTCM p. 200

Colonel Lane reported, "At 10 p.m. we were moved to the rear on another ridge, and before the morning of the 5th had completed rifle-pits of the length of two regimental fronts, having built during the night three times the length of our own front of rifle-pits and dug the trenches, the men not having slept or rested for upward of thirty-six hours." WOTH Vol 25 p. 765

May 5, 1863

For the Union Army 12th Corps the day passed pretty much as the day before. The Confederate Army did not attack. They men constructed evergreen bowers for shelter under the slope of their natural fortresses. For food they broiled pork on sticks and ate those with hardtack (soldiers' hard bread). At about 3 p.m. they received a heavy storm of thunder, lightning and rain. Generals Geary and Williams received orders to be ready to recross the Rappahannock soon after dark, but that theirs would be the last corps to move and it would probably be 10 or 11 p.m. when they would move. General Williams wrote: "It was darker than Erebus. We gathered around a big camp fire and piled on the wood, but the cold rain poured, and done up in our rubber coats and hats we looked most forlorn and felt quite so." Colonel Lane reported, "Rested in our rifle-pits until 10 p.m., when we were called under arms, and the brigade was formed in close column of regiments and remained under arms until daylight of the morning of May 6, when we marched to the United States Ford, and crossed, after which, by ordinary marches, we again reached Aquia Creek in two days and made camp." FTCM p. 201

Meanwhile near Fredericksburg to the east, Union General Sedgwick and the Union Army VI Corps withdrew across the Rappahannock River at Banks's Ford during the pre-dawn hours of **May 5**. When he learned that Sedgwick had retreated back over the river, Union Army of the Potomac Commander General Hooker felt he was out of options to save the campaign. He called a council of war and asked his corps commanders to vote about whether to stay and fight or to withdraw. Although a majority voted to fight, Hooker had enough, and on the night of **May 5–6**, he withdrew back across the river at U.S. Ford.

It was a difficult operation. Hooker and the artillery crossed first, followed by the infantry beginning at 6 a.m. on **May 6**. Meade's V Corps served as the rear guard. Rains caused the river to rise and threatened to break the pontoon bridges.

Union General Couch of II Corps was in command on the south bank after Hooker departed, but he was left with explicit orders not to continue the battle, which he had been tempted to do.

The surprise withdrawal frustrated Confederate General Lee's plan for one final attack against Chancellorsville. He had issued orders for his artillery to bombard the Union line in preparation for another assault, but by the time they were ready, Hooker and his men were gone.

The Union cavalry under Brig. Gen. George Stoneman, after a week of ineffectual raiding in central and southern Virginia in which they failed to attack any of the objectives

Hooker established, withdrew into Union lines east of Richmond—the peninsula north of the York River, across from Yorktown—on **May 7**, ending the campaign. History-maps.com: Battle of Chancellorsville

General Williams wrote to his daughter on **May 7**, "You will be startled to see that I am back to the old camping ground (at Stafford Courthouse). But so it is and sadly so. After ten days of great hardship, exposure, and privations we are back again with a diminished and dispirited army. We recrossed the Rappahannock yesterday morning, the whole army moving after midnight over two pontoon bridges. My division was the last to escape, except the rear guard… I am by no means cheerful, because I think this last battle has been the greatest of all bunglings in this war. I despair of ever accomplishing anything so long as generals are made as they have been." FTCM Pg 177-178

Lee, despite being outnumbered by a ratio of over two to one, arguably won his greatest victory of the war, sometimes described as his "perfect battle." But he paid a terrible price for it, taking more casualties than he had lost in any previous battle, including the Confederate defeat at the Battle of Antietam. With only 60,000 men engaged, he suffered 13,303 casualties (1,665 killed, 9,081 wounded, 2,018 missing), losing some 22% of his force in the campaign—men that the Confederacy, with its limited manpower, could not replace. Just as seriously, he lost his most aggressive field commander, Stonewall Jackson.

Brig. Gen. Elisha F. Paxton was the other Confederate general killed during the battle. After Confederate General Longstreet rejoined the main army, he was highly critical of Lee's strategy, saying that battles like Chancellorsville cost the Confederacy more men than it could afford to lose.

Of the 133,000 Union men engaged, 17,197 were casualties (1,606 killed, 9,672 wounded, 5,919 missing), a percentage much lower than Lee's, particularly considering that it includes 4,000 men of the XI Corps who were captured on **May 2**. When comparing only the killed and wounded, there were almost no differences between the Confederate and Federal losses at Chancellorsville. The Union lost three generals in the campaign – Maj. Gens. Hiram G. Berry and Amiel W. Whipple and Brig. Gen. Edmund Kirby. Geary's division, which included the 102nd NY Volunteers, lost 1209 men. The 102nd lost 90 killed, wounded and missing. Moses Whitbeck was not one of them. Colonel Stainrook, of the 109th Pennsylvania, who had previously commanded the brigade, was killed in gallant action leading his men. Chancellorsville Battle Civil War Campaign Casualties Killed. Excerpts taken from Thomaslegion.net: American civil War: Chancellorsville Campaign

May 9, 1863 The 102nd NY Volunteers are camped near Aquila Creek. Maybe there was some discussions between generals about what really happened on **May 3**, as Colonel Lane made a supplementary report:

> CAMP NEAR AQUIA CREEK, VA., May 9, 1863. CAPT.: I have the honor of making a supplementary report to you, more fully describing the movements of the left of the brigade at the time of receiving the order to fall back from the trenches on the morning of May 3. A short time before we received the order to fall back, say about five minutes, the lower end of the line in the rifle-pits nearest the enemy broke badly, leaving the trenches. Here, I believe, were the Seventy-eight New York Volunteers and about three companies of the One hundred and forty-ninth New York Volunteers. As they

neared the right of the One hundred and second, Capt. Mead, of Company K, One hundred and Second New York Volunteers, left the trench, and, running toward them, struck with the flat of his sword the nearest man, and endeavored to stop the movement. He was followed by the color-bearer, and the colors were placed about 20 yards from the trenches and to the rear of them, and the order given, "To the colors!" when Company K, on the left, and Company H, on the right, formed immediately, and the rest of the regiment extended the line left and right at right angles to the trenches, after which the officers rallied the retreating men of the lower regiment behind us. One lieutenant of the One hundred and forty-ninth, with his men, came and formed with us, asking for a place. I do not know his name. The regiment thus formed immediately opened fire on the rebels advancing on the trenches, and stopped and finally broke them. It was just after we had formed across the line that I received the order to fall back in good order. This was impossible to do at the time, from the proximity of the rebels. As soon as the enemy fell back slightly, our men cheered. We then fell back about 100 feet to the brush and abatis made by the Sixtieth. Capt. Mead asked Col. Redington if he would remain and support us if we went again in front of the screen, as the rebels were again advancing. He said he would, when we immediately formed again directly in front of the cross abatis of the Sixtieth, and again opened fire, part of the Sixtieth forming with us, stopping the advance of the rebels. All the men below us in the trenches having passed, I gave the order to fall back in good order, at which time the captain, lieutenant, and men of the Twelfth Georgia Volunteers came in from both sides, and we took them prisoners (as per first report).

Capt. Mead, of Company K, One hundred and second New York Volunteers, then personally told Col. Redington that the orders were to fall back, and the One hundred and second retired through the trenches toward the plank road. Arriving there, we formed in front of the battery at the brick house and on its left flank. Immediately after, the Sixtieth and some part of the One hundred and forty ninth New York Volunteers and other formed on our right, skirting the woods, to enfilade the rebels should they try to take the battery. While here, we were fired into with grape from a concealed battery of ours in the woods opposite our right flank, killing 1 man of the Sixtieth and threatening the whole. I immediately gave the command to fall back, and at the instance of Lieut.-Col. Dickinson, of Gen. Hooker's staff, we took up position No. 2, where we stayed, supporting that battery, and remained until it left and was replaced by another, and that again left. When the second battery was completely in rear of the line of battle formed on the right of the brick house, we rose from our cover, and as fast as possible joined our brigade. Very respectfully, your obedient servant, JAS. C. LANE Col., Cmdg. One hundred and second New York Volunteers. WOTR Vol 25 p. 766-767

Gettysburg Campaign June 11-July 24, 1863. After the Confederates' victory at Chancellorsville in May 1863, General Robert E. Lee's Army of Northern Virginia and the Union

Army of the Potomac, commanded by Maj. Gen. Joseph
Hooker, once again confronted each other across the Rap-
pahannock River near Fredericksburg. Lee's army slipped away
from Federal contact at Fredericksburg on **June 3, 1863**. The
largest predominantly cavalry battle of the war was fought at
Brandy Station on **June 9**. The Confederates crossed the Blue
Ridge Mountains and moved north through the Shenandoah
Valley, capturing the Union garrison at Winchester, in the Second Battle of Winchester, **June 13–15**. Crossing the Potomac
River, Lee's Second Corps advanced through Maryland and
Pennsylvania, reaching the Susquehanna River and threatening
the Pennsylvania state capital of Harrisburg. However, the
Army of the Potomac was in pursuit and had reached Fred-
erick, Maryland, before Lee realized his opponent had crossed
the Potomac. Lee moved quickly to concentrate his army
around the crossroads town of Gettysburg.

On **June 3, 1863**, Lee's army began to slip away north-
westerly from Fredericksburg, leaving A.P. Hill's Corps in
fortifications above Fredericksburg to cover the army's de-
parture, to protect Richmond from any Union incursion across
the Rappahannock, and to pursue the enemy if Hill thought it
advantageous. The 102nd NY Volunteers were in Aquia Landing, Virginia

June 9, 1863 The Battle of Brandy Station, also called the
Battle of Fleetwood Hill, was fought and was the largest predominantly cavalry engagement of the American Civil War, as
well as the largest ever to take place on American soil. It was
fought on, and around Brandy Station, Virginia, at the beginning of the Gettysburg Campaign by the Union cavalry
under Maj. Gen. Alfred Pleasonton against Maj. Gen. J.E.B.
Stuart's Confederate cavalry.

Union Commander Pleasonton launched a surprise dawn attack on Stuart's cavalry at Brandy Station. After an all-day fight in which fortunes changed repeatedly, the Federals retired without discovering Gen. Robert E. Lee's infantry camped near Culpeper. This battle marked the end of the Confederate cavalry's dominance in the East. From this point in the war, the Federal cavalry gained strength and confidence. Seeing the Civil War: Brandy Station VA

The Second Battle of Winchester was fought between **June 13 and June 15, 1863**, in Frederick County and Winchester, Virginia as part of the Gettysburg Campaign during the American Civil War. As Confederate Lieutenant General Richard S. Ewell moved down the Shenandoah Valley in the direction of Pennsylvania, his corps defeated the Union Army garrison commanded by Major General Robert H. Milroy, capturing Winchester and numerous Union prisoners. The 102nd NY Volunteers were still camped at Aquila Landing.

June 13, 1863 The Second Division was still at Aquia Landing. Even though they have been on the alert to move for some time, they have not left. General Geary wrote his wife Mary: "The facts are, we have been under marching orders for some time, to move at a moments notice, and the booming of cannon is constantly heard in our front, indicating that a general battle may at any time occur... I have very little doubt that we are on the eve of stirring events. Our army and that of Lee's are constantly manoevering and you no doubt observe the collisions that are occurring, as described in the newspapers. Tonight I am informed that some of the troops from Fortress Monroe will pass the Chickahomonie river toward Richmond. Also the troops from Suffolk will pass the Blackwater in the same direction. Thus you cannot fail to see that there is work at hand. Tomorrow morning two of my brigades will move from this place South-westward a few miles. I will know in a few days what part of the play my command is to perform." General Geary wrote nothing else until June 19th from Leesburg after two hard days of marching. APGTW p. 91

June 19, 1863 Now in Leesburg, Virginia, the 12th Corps of the Union Army had a very solemn military execution of three soldiers from the First Division that the entire Corps had to witness. General Geary wrote his wife: "the solemn spectacle of a military execution took place in our corps. Three men belonging to Gen. William's Division convicted as deserters were shot to death in the presence of the whole command. It was certainly a solemn scene, and one never to be forgotten." APGTW p. 93   One of the deserters had been gone a year and did not take the President's amnesty offer that expired in April.

June 20, 1863 The Second Division is still in Leesburg, and they spent the time strengthening their position. It was a calm day, and there was no sounds of guns according to General Williams FTCM p. 217 Leesburg, Virginia, was a prosperous southern town of about 1,700 at the outbreak of the Civil War. It was located strategically near the border of Maryland, just two miles south of the Potomac River. Leesburg was built in 1740 and is named for the Lee family, early leaders of the town and ancestors of Robert E. Lee. In the War of 1812, it was a asylum for important federal documents evacuated from Washington DC, and in the Civil War it changed hands many times.

June 21, 1863 The Second Division is still in Leesburg. About 10 miles south, a battle was culminating in Aldie as both sides had probed each other's position the last few days and

Union forces drove Stuart's Calvary away. General Geary wrote his wife, "We know not the moment we shall engage the enemy, he is undoubtedly before us. We have a pontoon bridge across the Potomac at Edwards Ferry. Our right rests on Ball's Bluff, thus connecting us with Maryland above Washington. We have also a pontoon bridge over Goose Creek on our left. Our extreme right covers the old Battle ground. I have just ascertained that the fighting before mentioned was between our calvary and that of Stewart [Stuart]. The enemy were beaten and driven several miles with considerable loss." APGTW p. 94

The confederates lost about 250 men and one cannon. The Corps spent the time in Leesburg according to General Williams: "Those [days] were devoted to putting the forts in good condition for the rebels and making several miles of rifle pits and breastworks." FTCM p. 222

June 23, 1863 The Second Division is still camped near Leesburg. Despite having to be on the defensive and building breastworks, they were enjoying their time there as much as possible because the people of Leesburg thought it was the Confederates who were originally coming, so they prepared all kinds of favors and foods for them. It had been a race between the Union Army and Confederate General Longstreet's Corps to see who could get there first. General Geary wrote his wife, "When our army was coming in, the people of Leesburg were taken completely by surprise. They were busy cooking and baking for their friends in the rebel army, who they were momentarily expecting... The people were crest fallen, but they try to make the best of it by sharing with us some of the delicacies prepared for other palates. APGTW p. 95

June 26, 1863 The second Division left Leesburg, marched across the river on pontoons at Edwards Ferry, went north toward Frederick, and camped at the mouth of the Monocacy Creek. The day was drizzling rain, and they marched 15 miles.

June 27, 1863 The Second Division marched north again, this time going through Point of Rocks at the Potomac River and reaching Knoxville. Point of Rocks had been an important crossroads of travel since American Indians established routes through the region. During the Civil War, troops from both sides frequently crossed the river and the towpath. Troops traded volleys across the water, skirmished in and near Point of Rocks, and Confederates attacked canal boats and trains, destroyed locks, and raided supply stores. On this day, they marched 18 miles through an abundance of cherry trees and although the day was cloudy the men were in good spirits. A change of orders came, and the Corps was directed to go toward Frederick. The 102nd NY Volunteers with about 250 men, as part of the Third Brigade led by General George S. Greene, crossed the Potomac River on a pontoon bridge and entered Maryland. They were marching north to meet the Army of Northern Virginia, which had already crossed into Pennsylvania. The Third Brigade was composed of five NY Regiments, the 60th, 78th, 102nd, 137th, 149th.

Also on this date, Union General Hooker is relieved of command and General George Meade is put in charge. General Geary liked General Hooker and wrote nothing about it. General Williams of the First Division, who sometimes was in charge of the Corps when the Commanding General was disabled or changed, wrote to his daughter, "I have said very little in my letters, but enough for you to guess that I had no confidence in Hooker after Chancellorsville. I can say now, that if we had had a commander of even ordinary merit at that place the army of Jackson would have been annihilated. I cannot conceive of greater imbecility and weakness than characterized that campaign from the moment Hooker reached Chancellorsville and took command." FTCM p. 221

June 28, 1863 The Second Division left Knoxville and marched to Frederick, Maryland. They passed through Jefferson and set up camp after a 15 mile march. While marching through the town of Frederick, the townspeople welcomed them with bands playing. That afternoon, news came to the men that General Hooker had been replaced and there was general satisfaction expressed by many. That night, they also received news that the next morning they were headed north toward Taneytown. Some men received new shoes and socks. APSL p. 7

June 29, 1863 The 12th Corps marched early, heading north from Frederick toward Taneytown but there was much traffic and delays, as the weather was cloudy and occasionally drizzling. They camped at Little Pine Creek after traveling 18 miles. They were close to the Pennsylvania state line, and the men were excited, as they knew that Lee's army was in their territory.

June 30, 1863 The Corps marched about 12 miles toward Littlestown, passing Taneytown. The First Division of the Corps led by General Alpheus Williams and his Brigade under General Candy was in the lead and pushed away some calvary from J.E.B. Stuarts' rebels. The local inhabitants were grateful and excited that the Army was there to protect them against the invading rebels. More supplies were given out to the soldiers, some received new uniforms that were badly needed. The soldiers had time to pitch their tents and cook meals while the afternoon sun was still out. APSL p. 8

Battle of Gettysburg July 1–3, 1863

The Battle of Gettysburg was fought **July 1–3, 1863**, in and around the town of Gettysburg, Pennsylvania. In the battle, Union Maj. Gen. George Meade's Army of the Potomac defeated Confederate General Robert E. Lee's Army of Northern Virginia, stopping Lee's invasion of the North. The three-day battle involved the largest number of casualties of the entire war and is often described as the war's turning point because of the Union's decisive victory.

Parts of the two armies met at Gettysburg on **July 1, 1863**, as Confederate General Lee concentrated his forces there, his objective being to fight the Union Army and destroy it. He believed his men were invincible. The area to the northwest of Gettysburg was defended initially by a Union cavalry division under Brig. Gen. John Buford, and soon reinforced with two corps of Union infantry. However, two larger Confederate corps assaulted them from the northwest and north, routing the Union forces, sending the defenders retreating through the streets of the town to the hills just to the south of town.

July 1, 1863 The 102nd NY Volunteers as part of the Union Army 12th Corps marched early about 5 a.m. and went 8 miles with their brigade on the Baltimore Turnpike from Littlestown, Pennsylvania, to Two Taverns. The men could hear some cannonading, so some soldiers climbed roofs or hills to get a look. ASPL p. 9 Soon they received orders to march toward the action, and they marched another 5 miles to a position on the extreme left of the army on Cemetery Ridge where the Union Army had been pushed. Geary's men left the Baltimore Pike and then formed battle lines in the field, as they moved forward. They engaged with rebel skirmishers that evening, but darkness ended the engagement. They stayed in that location that night, resting on their arms as it was called, until shortly after daylight of the next morning.

Captain Stegman wrote the report for the 102nd as Colonel Lane was injured: July 1, this regiment marched with the brigade from Two Taverns to a position on the extreme left of our army, then engaged with the enemy's skirmishers; were advanced, and the position occupied until shortly after daylight the next morning, July 2.

The division being moved to the right of the army, the One hundred and second New York was formed in line upon the side of a precipitous hill; the One hundred and forty-ninth New York upon the right, and the Sixtieth New York upon the left. Skirmishers and pickets from the First Corps occupied our front, but were relieved by detail. The men were ordered to build breastworks, and did so with the best material at hand—cord-wood and rock—making, however, a strong line. The Sixtieth and One hundred and forty-ninth New York Regt.'s extended their lines to connect with ours, thus forming a long and continuous breastwork. Artillery firing took place from our immediate rear upon the enemy, drawing a fire in reply, but doing no serious damage. This occurred about 4 p.m., and continued for about an hour, perhaps more. Shortly after 6 p.m. the regiment was moved by the right flank to the intrenchments occupied by the One hundred and forty-ninth New York, the men forming in single file, with intervening spaces of a foot or more. The men had scarcely taken this position when some sharp musketry firing took place, proving an advance of the enemy, and causing our pickets to retire. The Seventy-eighth New York was dispatched through our lines to their relief, bravely led by Lieut.-Col. Hammerstein. The blaze of fire which lighted up the darkness of the valley below us; the desperate charging yell and halloa of the rebel troops, convinced us of an immediate engagement. The men were cheered by their officers, continued to be on the alert, and to allow our pickets to pass.

The Seventy-eighth soon fell back in good order before the heavy columns of the foe, forming on the rear of our right wing, where they remained during the battle, relieving our men and in turn being relieved, fighting desperately and bravely. The pickets having crossed the breastworks, the whistling of the balls announced the advance of the enemy to close quarters. It was answered by volley after volley of the most destructive muskery from our regiment, being unceasing for two hours.

About 8 p.m. the right wing was re-enforced by the Sixth Wisconsin Regt., Wadsworth's division, First Army Corps. About 8.30 p.m. the left was re-enforced by the Forty-fifth New York, Eleventh Corps, which occupied the position until the firing ceased. The One hundred and second New York never left its position, nor did one man flinch from his full duty. The firing ceased along the line about 9.30 p.m. During the night, about 1 a.m. and again at 2 a.m., volleys were delivered by both sides. WOTR v. 27 pt. 1 p. 864-865

July 2, 1863 The Second Division was awakened that morning by the sound of the Union Army 3rd Corps coming up the Emmitsburg Road to occupy the ground on the Union left. The 102nd NY Volunteers with Private Moses Whitbeck were moved to the right of the Union Army on Culp's Hill, about a mile and a half, as Part of the Union Army 12 Corps under Union Major General Slocum, about 10,000 troops. Union Brigadier General John W. Geary Commanded the Second Division, and the 102nd NY was in the Third Brigade

These pictures show the monument where the 102nd fought. This is at the top of Culp's Hill and the hill slopes down to the right.

This picture is from the top of Culp's Hill looking down to where the Confederates were. The land is much more foliated than it probably was back in 1863. Also the trees are taller than they were back then. It was explained to me when we were there that trees back in that time were cut down when they reached the circumference of the size of a fence post.

under the oldest General in the Union Army, Brigadier General George S. Greene. The 102nd was led by Captain James C. Lane. It was a long hot summer day, and Greene's Brigade of about 1500 NY infantry was left alone on Culp's Hill while the rest of the Division had been sent to the Union Left to bolster the defenses against Confederate General Longstreet's attack on the Union Left. For those remaining on the slopes of Culp's Hill, the move was met with justifiable concern. Brigadier General Alpheus Williams wrote, "I received an order from General Slocum to send a division to reinforce the left of our

This picture was taken down by Rock Creek where the Confederates were. Could you image having to climb up that hill to attack the Union position all while under fire? Many men died here and the Confederates never took the hill. Many cried when they were ordered to retreat.

army, which was reported as being hard pressed. Lockwood and the 1st Division being nearest to the line of march, I ordered both to move out around the base of Power's Hill…and I immediately reported to General Slocum that I had great fear that the rebels would seize our line on the right the moment we left it; that I had ordered Geary to cover the whole line and thought we could not safely spare more of our troops from that position. General Slocum said the call for troops was urgent for all the troops he could spare, but he approved of my suggestions that at least one division was necessary to hold our entrenchments. With this understanding, I joined the head of the reinforcing column." Culp's Hill at Gettysburg, John M. Archer p. 52-53

Greene's Brigade was erecting breastworks on Culp's Hill during the day and moving their lines to cover the spaces previously occupied by about 6400 men, waiting for the Confederate's to attack. Captain Stegman of the 102nd NY observed that to cover this distance there was a very thin line, the men being fully a foot apart or more, in single rank. They had scarcely taken this position when they were met by a rebel advance. The 78th NY Regiment came to their aid and together both regiments took turns relieving the other while the Confederates made a desperate attempt to take the hill.

About 7 p.m., as it was starting to get dark (no daylight savings time back then), Confederate Major General Edward Johnson launched an attack on Culp's Hill with about 4500 infantry. On the far Union right on lower Culp's Hill sat the largest regiment of the Third Brigade, the 137th New York under Colonel David Ireland with 450 men. The Confederates had about 2300 men to face them. The initial frontal attack by the Confederates was repulsed. But the Confederates sent about 1000 troops further over to the Union right to try to flank the 137th NY. Colonel Ireland sent two companies of men (about 70 soldiers) to try to stop the Confederate flanking maneuver. It didn't work, and Ireland's regiment was forced to fall back up to the top of Culp's Hill to the recently made breastworks. The Confederates pursued, but the 137th was able to hold them off long enough for Union General Greene to send reinforcements.

The 102nd NY Volunteers were in the middle of Greene's defense on Culp's Hill and faced part of Jones's Confederate brigade from Virginia and part of Nicholl's Louisiana Brigade. Casualties for the regiment were relatively low due to the breastworks General Greene insisted they build. But the men

This picture is from the side of Culp's Hill or what would be called the flank. This shows where the 102nd was located and the slope going down to Rock Creek where the Confederates came up from is to the right.

were running low on ammunition.

General Greene then did something very unusual – he ordered his men to make a bayonet charge in the dark that pushed the Confederates back to lower Culp's Hill. The move by General Greene saved the day, as had the Union been flanked that night, disaster would have struck and changed the battle. The order in which Johnson's Confederate brigades reached Culp's Hill is debatable; what is clear is that the rebel advance was halted abruptly only yards from their goal. The men of the 12th Corps under Union General Alpheus William's leadership were now coming back from the Union left late at night, and the men hunkered down at the top of

Culp's Hill and awaited daylight. Both sides were reinforced during the night.

Colonel Lane of the 102nd NY Volunteers was wounded that evening about 9 p.m. Captain Lewis Stegman took his place, and in his report after the battle he wrote, "The 102nd NY never left its position, nor did one man flinch from his full duty…Too much credit cannot be given to the officers and men of the regiment for their unflinching courage and devotion; when their ammunition was expended, with determined spirits they awaited the enemy's onset with fixed bayonets. True and trusty, they have added renewed luster to the bright name already borne, so hardy won on many a desperately contested field… Where every officer acted conspicuously, bravely, and courageously, I can scarcely find one action more creditable than any other in any one. All acted most nobly and heroically." WOTR Vol 27 pt. 1 p. 865

Greene's New York Brigade repelled four different Confederate charges. The 102nd NY Volunteers and the other regiments were still holding their ground, even Ireland's 137th Regiment, which took heavy casualties. Greene's troops held the hill alone for several hours, barely hanging on with some help of several hundred reinforcements from the First and Eleventh Corps. The 14th Brooklyn relieved the 137th. About 10 p.m., Barnum's 149th NY Regiment, which had the biggest hit in casualties, was relieved. Now the men could clean their muskets and replenish their ammo. The rest of the men spent the night in their works preparing for the next day's battle. Both sides were reinforced that night.

July 3, 1863 Confederate Major General Edward Johnson's Division was now at least 9000 men strong. Union General Slocum, with about 9,800 men, planned to attack at first light.

At 4:30 a.m., Slocum's 12 Corps started the longest sustained action during the entire battle, about 7 hours of non-stop fighting. About 23,000 men were involved in this part of the battle, more than the number of soldiers engaged during Picket's charge later in the day. News came about 10:30 a.m. that Confederate General Longstreet's attack on the Union left would soon begin. Now the Confederates under General Johnson would renew their push against Culp's Hill. After seven hours of fighting to try to take Culp's Hill, the Confederates had lost about a third of their men. The men of the 102nd NY fought for more than six hours alone until finally relieved by the 150th NY, who they had to relieve again only 20 minutes later. About an hour later, the 150th NY relieved the 102nd NY as the fighting was intense. The 102nd had scarcely reorganized when the 1st Maryland was pushed from their trenches, and the 102nd NY were ordered to fill the gap where they fought for several more hours until relieved by the 60th NY Volunteers.

About noon, Confederate Lieutenant General Richard Ewell had decided that his men could not take Culp's Hill, and they retreated. This part of the battle of Gettysburg was over; the men of the 12th Corps would remain atop Culp's Hill and not participate further. Other than sporadic clashes in the dark along the skirmish line, the struggle for Culp's Hill ended at sunset. The coming darkness put an end to the fighting, and the men of the Third Brigade and the 102nd NY were worn out with hunger and the day's events and fell asleep. A soldier from the 149th NY wrote, "The men tried to keep awake, but it was impossible on account of their excessive fatigue…the strain on the nerves by the concussion of firearms during the day wearied the body beyond description."

After the failure of the afternoon assault, Confederate General Lee issued orders for Ewell's Corps to retire that night to the army's new line west of Gettysburg. Twenty-four hours after capturing the lower summit of Culp's Hill, Johnson's brigades recrossed Rock Creek and silently withdrew to the west. Many of his men were in tears.

"Pop" Greene's Brigade and the rest of the Division were thankful he had the wisdom to insist on the breastworks being made. Without those defenses, the brigade could have never held off the rebels. The men of the 102nd NY Volunteers helped hold the critically important Culp's Hill. They didn't even know the name of the place they fought until after the battle was well over.

July 4, 1863 Early morning probes by Union pickets confirmed that Confederate General Johnson's troops had without doubt withdrawn from the Culp's Hill and Rock Creek area. The word spread quickly among the 12th Corps and the 102nd NY Volunteers. Sgt. Sam Lusk of the 149th NY Volunteer Infantry would write, "The men were awakened at daybreak to a glorious Fourth of July, with the salutation, 'The enemy has skedaddled, and we are masters of the field.'" Because they were occupied by the fight for their survival for the last 36 hours, they were just starting to understand their surroundings and what transpired. Lusk wrote "we were a hard looking set of fellows when the battle was over...our faces were as black as coal. Our clothes were covered with blood and dirt. Some places in the trenches were saturated with human blood." The slaughter on Culp's Hill was appalling, even for the veterans of Greene's Division. Sgt L.R. Coy of the 149th NY wrote, "It made the men sick in body and mind... The havoc in the Union lines was terrible, but among the enemy it was even more so… in a spot not 12 feet square I saw 8 dead, but I cannot describe what I saw, it was too horrid, truly I thought as I passed over the field none but demons can delight in war." The losses on Culp's Hill were heavy, of the approximately 6400 soldiers engaged in Confederate General Johnson's division, the struggle for Culp's Hill left more than 2,000 men killed, wounded and missing. Although Gettysburg was also costly for the Army of the Potomac, the defenses on Culp's Hill had proven to be worth it. Of the 9,800 engaged, casualties for the entire 12th Corps were almost 1,100 men. Some of the hard-fought regiments of the 12th Corps, such as Greene's 137th NY and Colgrove's 27th Indiana and 2nd Massachusetts, would bear the brunt of these losses. The 102nd NY brought 248 men to the field, losing four killed, 17 wounded, and eight missing. Colonel Lane was wounded on **July 2**, and his command was taken by Captain Lewis R. Stegman. The regiment also lost Captain John Mead and Adjutant J. Virgil Upham. Praise God that Private Moses Whitbeck survived the Battle, or I would not be here today writing this book.

Pursuit of Lee to Manassas Gap, Va., July 5–24, 1863

July 5, 1863 The 102nd NY Volunteers and the Third Brigade left Gettysburg and marched to Littlestown, which is about 10 miles Southeast of Gettysburg on the Baltimore Pike.

Littlestown had been the site where on June 29th General Kilpatrick's Division of the Union Calvary bivouacked for the night. Kilpatrick and General George Custer both lodged at the Barker House. The next morning in Union Mills Maryland, General Jeb Stuart received word from his scouts that a large force of Union calvary had been spotted in the vicinity of

Littlestown, and a local teenager, a 16-year-old Herbert Shriver, volunteered to guide the Confederates on a detour around Littlestown by way of Hanover. Stuart did not know that Kilpatrick's forces were already on the move to Hanover; General Kilpatrick was also unaware of Stuart's detour to Hanover, and both were quite surprised when they clashed in what was known as the Battle of Hanover.

After learning that Stuart's Troops were defeated at Hanover, General Slocum's Corps of 13,000 infantry entered Littlestown that evening and was dispatched to Gettysburg the next day. General Sedgewick's Sixth Army Corp of 15,000 also passed through Littlestown on their way to the Battle of Gettysburg. Now the 12th Corps was passing through Littlestown going the other direction and following the retreat of Lee's army, but they would stay there until July 7.

July 7, 1863 The Second Division resumed their march at 4 a.m. and passed through Taneytown, Middleburg and Woodsborough, marching 29 miles they camped near Walkersville, Maryland. Taneytown was where General Meade had his HQ at the Shrunk Farm from June 30th until the night of July 1st. Walkersville was south and slightly west of Gettysburg.

July 8, 1863 The Second Division left their camp near Walkersville at 5 a.m. and marched through Frederick and over the Catoctin Range and halted for the night a half a mile beyond Jefferson. General Geary wrote his wife, "Lee has been whipped, and routed, and we are now in hot pursuit. Tonight we expect to reach Crampton's Pass. There may be some hard fighting yet." Union General-in-Chief Henry Halleck began exhorting General Meade to move more rapidly and to attack Lee when the Confederate army would become divided during its crossing of the Potomac River.

July 9, 1863 The Second Division left camp near Jefferson at 5 a.m. and marched through Crampton's Pass and Burkittsville to Rohrersville and camped at 11 a.m. Union General Meade began the movement to the west of South Mountain, which he described as being "on a line from Boonsboro toward the centre of the line from Hagerstown to Williamsport, my left flank looking to the river, and my right toward the mountains." Although he had surmised Lee's next objective, Meade was still unsure of his opponent's exact whereabouts and could not exclude the possibility of being attacked by the Confederate army. As he conveyed to Halleck, "It is with the greatest difficulty that I can obtain any reliable intelligence of the enemy."

July 10, 1863 The 12th Corps left Rohrersville at 5 a.m., passed through Keedysville, and reached Bakersville at 11 a.m. Calvary pickets of the Confederates left the area because of the Union advance. The Union formed a line of battle with the First Division on the left and II Corps on the right. Some light breastworks were made, and heavy pickets advanced, then they bivouacked that night.

July 11, 1863 The Second Division of the Union Army 12th Corps advanced to Fair Play and formed in line of battle and sent skirmishers forward toward calvary pickets in their front but stayed in this position all night.

July 12, 1863 The Second Division, which included the 102nd NY Volunteers as members of the Third Brigade, advanced a line of pickets to a more elevated ridge, resulting in some slight skirmishing until the rebels fell back. During the night, they changed their position to the right, then withdrawing to a ridge to their left and joining the line that was to the left of the First Division.

July 13, 1863 The Corps worked all day and covered their entire front with extensive breastworks and numerous traverses to protect their flanks all while under fire from Confederate pickets. At 5 p.m., the pickets from the Second Division were ordered to advance until they met those of the enemy. They found them under cover of belts of woods about a quarter of a mile in their front, and lively skirmishing ensued, but the unit received no casualties. At dusk, the Union went back to their original line.

July 14, 1863 The 12th Corps Second Division that included Private Moses Whitbeck and the 102nd NY Volunteers, remained entrenched and under arms to support the First Division which advanced to the front. Here the First Division discovered that there was no opposition in front of them and General Geary of the Second Division ordered his skirmishers, the 28th Pennsylvania and Seventh Ohio Regiments to do reconnaissance toward Downsville. They discovered some enemy works there were vacated and returned with several prisoners.

July 15, 1863 The Second Division marched at 6 a.m. and camped within 4 miles of Harpers Ferry and got there at 4 p.m.

July 16, 1863 Resuming their march from the previous day they made it to Pleasant Valley by 7 a.m. and set up camp. They had marched 101 miles and suffered much from the excessive heat of that summer. Here, they would stay for several days to rest and to replenish supplies.

July 17, 1863 The 12th Corps is in Pleasant Valley, Maryland. They are preparing for an extensive march. General Geary wrote his wife, "We are now refitting the clothing and equipments of the command which will be completed tomorrow. We are under marching orders to have 3 days cooked Rations in haversacks & 3 days in wagons and there is consequently an extensive march before us. The result of the war seems no longer doubtful, and every thing in a military point of view seems more cheering than ever heretofore, the beginning of the end seems visible." APGTW p. 101

Many soldiers received new uniforms that were badly needed since the Battle of Gettysburg.

July 19, 1863 The Second Division left Pleasant Valley at 5 a.m. and leading the corps they marched through Harpers Ferry, crossing the Potomac and Shenandoah Rivers on pontoon bridges, and passed up Piney Run Valley to near Hillsborough where they camped. General Williams and the First Division were behind, and he wrote his daughters, "The guerrillas troubled the advance considerably, coming down from the passes in the Blue Ridge. We sent out patrolling parties, and as some firing had been done from houses, we arrested several citizens and copying Lee's example we gathered up all the horses and cattle we could lay our hands on. Our first camp was in the vicinity of the pass through 'Short Mountains' near Hillsboro. From this point we went last winter through Hillsboro to Leesburg. Our present march will be toward Snickersville." FTCM p. 237

July 20, 1863 The Second Division marched at 6 a.m. and went through Woodgrove to Snickersville where they stayed until the 23rd with the First Division to guard Snicker's Gap. The road there was awful and troops had to find alternate paths through fields and woods, the day was hot and humid. General Williams of the First Division wrote his daughters, "You may remember that I marched my division up to the pass on the other side just before the battle of Winchester, and was obliged to go back and go through Jackson up the Shenandoah Valley. Years ago, I passed over this same route from Berryville to Washington. It was before the days of pikes in this section. I found the most execrable roads, such as we came over yesterday; roads in which small mountain streams find their courses, making an alteration of rocks and mud as the nature of the soil changes. Roads are never repaired in this state, except the pikes." FTCM p. 238

July 21, 1863 There is some disagreement between the different Generals Geary and Williams about how long they stayed in Snickersville, but General Geary appears to be wrong when he said just one day. But, they did have some slight skirmishing there with guerilla bands and took a few prisoners and killed some rebels but each General painted the action there in a little different light. Both generals wrote interesting accounts of their trip with General Slocum that day up to the top of the Blue Ridge Mountains where they could see the rebels below in the Shenandoah Valley.

July 23, 1863 The Second Division marched at 5 a.m. and went to Paris, Virginia. But, the Second Brigade under Colonel Cobham was sent to Ashby's Gap, to guard it, relieving a brigade from the II Corps. At 4 p.m. they were ordered to march again, taking the mountain road and passed through Scuffletown to near Markham Station, where they came upon the wagon trains of the Second Corps. The day's march was 23 miles. The First Division was sent in a different direction as well. They went to Piedmont and Markham's Station on Manassas Gap.

July 24, 1863 The Second Division left early, at 3 a.m., and marched through Markham to Linden and got there at 8 a.m. They had marched 30 miles in the last 27 hours. They expected a battle but got only a skirmish. General Geary wrote his wife, "we reached there about 8 o'clock the next morning, having made a forced march of 30 miles under the expectancy of a general battle at that place. The fight was soon over for the enemy skedaddled. Several hundred of them were killed and wounded. We captured several hundred of the Pennsylvania cattle and horses, also a large number of their sheep and hogs." APGTW Pg 104   They remained under arms until noon, when they resumed marching. They passed Markham and camped at Piedmont after having marched that day 22 miles.

July 25, 1863 The Division was on the march again early, this time leaving at 4 a.m. and they marched through Rectortown and White Plains to Thoroughfare Gap, 16 miles. The roads were in very bad condition and the weather was oppressively warm.

July 26, 1863 The Division marched at daylight by way of Greenwich and Catlett's Station to near Warrenton Junction, 22 miles and set up camp. They were reunited with the First Division who had previously arrived by way of Manassas Gap. Here they remained until **July 31**, refitting the command. Since Leaving Pleasant Valley they had marched 103 miles, making a total of 204 miles from Gettysburg.

July 27, 1863 The 12th Corps was in Catlett's Station having just marched hard again and now getting some much needed rest. General Williams wrote his daughters, "I will continue my journal from this point if we stay here, as I hope, a day or so. My men greatly need rest. They have been marched now for nearly six weeks, much exposed, up nights, and always on the march as early as 2:30 or 3 o'clock. It surprises me, often, how they stand it, carrying, as they do, over sixty pounds weight and marching over roads of the roughest and rockiest kind, But they do it, day after day, in the old regiments without a single man falling out. At the end of the two days' march of forty-five miles over the hardest roads, of the old regiments three men was the highest number reported as stragglers, and it is very probable that they were sore footed or fell asleep on a halt, without waking." FTCM p. 246

July 30, 1863 The 12th Corps remained in Catlett's station for three days but received orders late that night to be ready to march first thing the next morning to Kelly's Ford.

July 31, 1863 The Corps marched in the early morning with the Second Division in the lead of the corps and they went by way of Elk River and Morrisville to Kelly's Ford, a march of 20 miles in very hot weather. The division skirmished with rebel calvary and drove them off. They bivouacked that night close to the river. The 102nd stayed near Ellis' Ford.

Duty on line of the Rappahannock until September 1863 What this means when you research about any Civil War unit and you read about what they did, when it says duty on line, they are talking about how the Union was on one side of the Rappahannock and the Confederates were on the other side. The regiments of men were constantly assigned duties such as: patrols; skirmishing, which meant spreading out along the riverbank and trees and shooting at each other; pickets, which meant men in groups scouting and skirmishing. This would become a large way they fought in the Tennessee and Georgia area because of the terrain being woody and with hills.

Also, after the Battle of Gettysburg was over, Union General Meade was cautious and lead inconclusive operations. This led to what was basically a stalemate along the Rappahannock River. Also in the Union Army, soldiers' two-year enlistment was coming to an end for many, and a good number of them thought they could leave the army just like that, but that wasn't necessarily the case. Yes, they could leave, but the pressure to re-enlist was large and often bonuses or encouraging offers were given by the feds or the state. Conscription and enlistment was a problem with both sides during the war and much has been written about it.

August 1, 1863 The Second Division was relieved by the First Division, which deployed six regiments as skirmishers up and down the river to drive away any rebel calvary that was still in the area. General Williams of the First Division wrote about the Kelly's who owned the Ford, specifically how they were rebels and slave owners, and since the older Kelly had sent his slaves south for protection, the Division looted his hay and grain and took all his horses and cattle. Since General Geary did not write about that, I assume that the Second Division and the 102nd NY Volunteers were not involved. We do know from Union war records that Moses Whitbeck was promoted to corporal on this day.

August 2, 1863 The Second Division marched to what General Geary called the "Camp in the wilderness Near Ellis Ford, Rappahannock River." He also said that it was the hottest weather he had ever seen. (This was something that the two generals, Williams and Geary, actually agreed upon) Their location was close to the meeting of the Rapidan and the Rappahannock rivers. General Geary wrote, "on the opposite side we can see two regiments of rebel Calvary and one Battery of artillery. The greater part of Lee's army is supposed to be massed at Culpepper, and a great battle would occur at any time if we attempted to cross over. Should the enemy remain quiet, I do not think we will move much until we are re-inforced by conscripts and otherwise. This is a very poor country. The inhabitants are in a starving condition. The only things that do grow here to any extent are whortleberries, blackberries, and snakes." APGTW p. 105

The 12th Corps now had pickets covering 7-8 miles of the Rappahannock River. General Williams of the First Division wrote his daughters, "We have been figuring up our month's work to make out the monthly report and find that we have marched, last month, eighteen days; engaged in skirmishing and battles five; in camp, but mostly under arms or packed for march, eight days. Since leaving Stafford Courthouse on June 13th we have marched over 440 miles, to say nothing of entrenching work, side marches, and small movements. Besides this, we are often up until 11 or 12, and must be around again by 3 o'clock and our men have heavy picket duty which keeps 300 to 400 without sleep after long marches. None but hardened troops could stand this. Their cheerfulness under it is wonderful." FTMC p. 250

August 3, 1863 The 12th Corps of the Union Army is at a camp near Ellis' Ford on the Rappahannock River. Here they will stay until late September. The weather was extremely hot and humid, and there was a heavy rain every afternoon. The men were in constant skirmish mode now and always on patrol. The people of Virginia were already weary of war, having much of it in their territory. General Geary wrote his wife, "The people here are very tired of the war, and would make peace today if they could induce their leaders to do so. If the leaders of the rebellion were but half as tired as the people, there would be peace in ten days." APGTW p. 106

(WOTR Volume 27 ends on **August 3** and Volume 29 picks off after that but there are no WOTR reports for the 12th Corps in Vol 29. In other words, there is not much in the official reports.

August 6, 1863 President Lincoln back on **July 15** issued a proclamation appointing this day as a day of national thanksgiving of praise and prayers for the recent victories of Gettysburg and Vicksburg. Some Units celebrated it while others did not. General Williams wrote his daughters, "National Thanksgiving day, which don't seem to be very generally observed. The regiment directly in my front had some kind of service toward evening. I heard a man holding forth in regular Methodistical roar, and psalm singing and hallooing prayers. There was a considerable confusion of tongues, for the adjoining regiments were singing patriotic songs with uproarious choruses, and the drums of other regiments were beating the adjutant's call for dress parade." FTCM p. 252

August 7, 1863 The Second Division is in camp near Ellis' Ford, Virginia. General Geary reported that the weather was hot but it rained every afternoon. The men are still busy on patrols, pickets and skirmish lines.

August 9, 1863 This was the 1 year anniversary of the Battle of Cedar Mountain. General Geary wrote, "This is also the ninth day of August, the anniversary of the battle of Cedar Mountain. The cycle of one most eventful year had passed away since that awful day of carnage and blood. How many who escaped that day have laid down their lives upon other fields perhaps more sanguinary but certainly not contested with more determined courage or unconquerable valor." APGTW p.108

In the Battle of Cedar Mountain, the 102nd lost Captains Julius Spring and Arthur Cavanaugh and 21 enlisted men killed or mortally wounded, Lieutenant Colonel Avery and six other officers and 70 enlisted men wounded, and one officer and 14 men missing.

August 22, 1863 The 12th Corps is still camped near Ellis Ford, Virginia. August had been particularly hot and humid, or "sultry," as they called it back then. General Geary wrote his wife, "A word or two this sultry morning will not be amiss. It is now over 4 weeks since we have had the honor to lead the van, or in other words, my "White Star" Division (called that because the Division Flag was a blue flag with a large white star on it) has occupied the most advance position of the army, and our pickets are constantly in view of those of the enemy. We are also on the left flank of the Infantry and are only flanked eastward by a portion of the calvary. Deserters from the enemy frequently come into my hands and we obtain much information concerning the operation of Lee's army." APGTW p. 109

September 1863 An important note is that the two armies faced each other along the Rappahannock River and there was not much movement. Both sides were still recovering from the Battle of Gettysburg in July, reorganizing and trying to figure out their next move. General Williams and General Geary wrote in their letters about the stalemate and that Confederate General Lee could attack at any time. General Williams in particular wrote about his dwindling numbers and the problems associated with that.

September 8, 1863 The Union Army 12th Corps was still on duty at the Rappahannock River. This meant that many men were on guard, watching the enemy and patrolling. General Geary wrote his wife, "There is considerable movement among the rebels this morning. Whether they intend to fall back, or to attack us or to make another invasion of Pennsylvania , we cannot tell-any one of which is probable and possible." APGTW p. 112

September 11, 1863 The Union Army 12 Corps is still on duty at the Rappahannock River, most of the days spent there seem to have some kind of skirmish as the calvary were very active probing the other side to ascertain position and strength. General Geary wrote his wife, "Every thing here bears the usual monotony of camp life. I am constantly watching with eager eyes every movement of the enemy. This morning there was quite a spirited cannonade some distance on our right, but I have been unable to tell what it was, probably a cavalry skirmish." APGTW p. 112-113 Turns out in the WOTR itinerary (Vol 29, p. 3) there was action at Middleburg that day.

September 13, 1863 The day was windy with a cold rain. It cleared up a little around noon, and the Second Division of the Union Army 12th Corps executed two soldiers for the crime of desertion. The entire Division, including Corporal Moses Whitbeck and the 102nd NY volunteers in the Third Brigade, were present as the men were shot to death by muskets. There was also a soldier shot for desertion in the First Division that General Williams wrote about. General Geary wrote that in all the army that day they shot 16 and that desertion was no longer to be tolerated. General Williams wrote that he had 20 more "conscripts" who would probably meet the same fate. General Slocum resumed command of the 12th Corps, and General Williams went back to his Division.

September 14, 1863 The Union Army is still on duty at the Rappahannock River, but they are starting to probe a little further with calvary, and as we can see looking back from history, they were getting ready for a short push to the Rapidan River as evidenced by the reconnaissance by Union calvary **September 13-17** toward Culpepper Courthouse, where the

Confederate Army was known to be.

September 16, 1863 The Union Army forces crossed the Rappahannock River and took up a position around Culpepper Courthouse where the Confederate army had been. The Second Division drove toward the Rapidan and Raccoon Ford. They stopped near Stevensburg.

September 17, 1863 The 12th Corps marched from Stevensburg to Racoon Ford on the Rapidan River. The rebels were on the other side and had fallen back to there. It was deemed they were in too much force for the Corps to cross. General Slocum decided to withdraw the 11th and 12th Corps from in front of the rebels as they would soon be transferred.

September 19, 1863 Somewhere behind Union lines near the Rapidan River, General Geary wrote his wife, "Since the rain of the 'equinoctial storm' the weather has become quite cool. So much so that it is unpleasantly cold. The enemy seems quite uneasy in our front. They are busy throwing up fortifications on almost every spot of available ground, and seemed determined to dispute the passage of the river, by us, under any circumstances." APGTW p. 114-115

The recent rains had made the roads mostly unpassable with mud, so it might be a few days until they move.

September 20, 1863 The 12th Corps was moved to an open plain further in the rear of the front. The next few days the men were kept packed to move and with rations in their knapsacks for eight days march away from their wagons.

Movement to Bridgeport, Ala., September 24-October 3, 1863

September 24, 1863 Because of the Union loss at the Battle of Chickamauga on September 18-19, The Union Army transferred the 11th and 12th Army Corps to the Tennessee area, and this would result in them eventually being combined into the 20th Corps. Secretary of War Edwin M. Stanton had arranged with the railroad presidents to transport 16,000 soldiers, 20,000 people in all, 1,233 miles through Union-held territory over the Appalachians and the Ohio River. The 102nd NY Regiment as part of the Third Brigade of the Second Division, of the 12th Corps were part of that movement.

After the Battle of Chickamauga, Union forces under Maj. Gen. William Rosecrans had retreated to Chattanooga, Tennessee. Confederate Gen. Braxton Bragg's Army of Tennessee then besieged the city, threatening to starve the Union forces into surrender. General Bragg's troops occupied Missionary Ridge and Lookout Mountain, both of which had excellent views of the city, the river, and the Union's supply line. The only supply line that was not controlled by the Confederates was a poor condition, tortuous route nearly 60 miles long over Walden's Ridge from Bridgeport, Alabama. Heavy rains began to fall in late September, washing away lengthy stretches of the mountain roads.

Confederate troops launched raids on all supply wagons heading toward Chattanooga, which made it necessary for the Union to find another way to feed their troops and get out of this lousy military situation. Part of that solution was to send the Union Army 11th and 12th Corps from The Hudson Valley to Chattanooga, and the other part was to place General Ulysses S. Grant in command. Because of his victory at Vicksburg, General Grant was now in command of the whole Western Theatre, and he would end up replacing General Rosecrans with General Thomas.

General Geary and his Second Division of the Union Army

12th Corps, which included the 102nd NY Volunteer Regiment under Colonel James C. Lane and with Corporal Moses Whitbeck, moved to the Chattanooga area with General Hooker's other troops to assist Grant in the campaign against Bragg's Confederates situated on Lookout Mountain and Missionary Ridge.

September 25, 1863 General Geary did not write many details about the trip from Virginia to Tennessee. But, General Williams wrote that they went to Bealton Station north of the Rappahannock and left their wagons and caught a train to Washington. General Williams had dinner with General Sherman. Most of the details of the trip that General Williams wrote about in a letter dated **October 5** to his daughters was unfortunately lost. FTCM P.265

October 1, 1863 The Second Division was in Belle Air, Ohio, and on the way to Cincinnati. General Geary described the journey to his wife, "We took cars at Bealeton Station near Rappahannock, and came through Washington, thence via Relay House, Harper's Ferry by the Baltimore and Ohio Railroad to this place. By tomorrow night we expect to be at Louisville and will take cars thence to Nashville, and thence by Rail to Chattanooga. The whole distance to be travelled will be about one thousand miles." APGTW p. 118

October 2, 1863 The Third Brigade of the 12th Corps Second Division, which included the 102nd NY Volunteers, were transported by rail through Columbus, Dayton, Indianapolis, Louisville and then Nashville to Murtreesborough, Tennessee, arriving there on **October 6**. They would remain there until **October 23**.

October 6, 1863 The Union Army 12th Corps, after a delay because the rebels had destroyed a railroad bridge south of

Murfreesborough, finally reached their destination of Murfreesborough late that evening. The town was already full, and the men probably had to camp somewhere on the outskirts. They had traveled over 1200 miles in two weeks.

October 8, 1863 The Second Division was now in Murfreesborough near the train depot. It is obvious from both General Geary and General Williams's books that the 12th Corps got to Nashville by way of Louisville and then on to Murfreesborough. From the Second Division the 28th Pennsylvania and the 66th Ohio regiments had a skirmish with the rebels and were able to push them away, overcoming a burned bridge on the railroad at Stone's River.

October 9, 1863 There was a rash of skirmishes caused by Confederate General Joseph Wheeler between **September 27 and October 17**. This day, small battles were fought at a railroad tunnel near Cowan, at Elk River, and at Sugar Creek, all in Tennessee. General Geary reported that 312 prisoners were sent to him in Murfreesborough, where he was in command. The 102nd NY Regiment was on duty as part of the Third Brigade.

October 12, 1863 The 12th Corps Second Division was still in Murfreesborough. General Geary wrote his wife a lengthy description of the area, basically praising it although War had worn it out already. He wrote his wife, "this is one of the best places in the world for the variety of its productions and for its unsurpassed abundance, but alas war has crushed its fairness and beauty, and caused it almost literally to return to its primitive wilderness." APGTW p. 124 The city was a hub with eight turnpikes coming together there and the Division, which included Corporal Moses Whitbeck and the 102nd NY Volunteers, were on guard duty. The town had a fine courthouse, five churches, and a couple of school houses and was generally a nice town.

General Williams also wrote a stirring account about the difficulty of the Union position. They had over 300 miles of railroad that was not only crossed every few miles by wide streams and valleys running through and across high mountains but was mostly in Confederate hands. So instead they tried to use a 30-mile mountain pass with bad roads to get supplies over. He surmised that before any real progress could be made that it was necessary that the rebels were driven off and the railroad opened up its whole length, but General Williams realized that was a difficult proposition.

On October 18, 1863 Union General Grant removed General Rosecrans from command of the Army of the Cumberland and replaced him with Major General George Henry Thomas. General Thomas was a Southerner who had stayed loyal to the Union. As the Civil War progressed, he won the affection of Union soldiers serving under him as a "soldier's soldier," they took to affectionately referring to General Thomas as "Pap Thomas." His strong defense at the Battle of Chickamauga in 1863 saved the Union Army from being completely routed, earning him his most famous nickname, "the Rock of Chickamauga."

October 19, 1863 The Second Division was still in Murfreesborough. General Geary wrote his wife about the coming change of command with Grant taking over and Replacing General Rosecrans with General Thomas. Then, the Division was given orders to move and went to Stevenson on the Nashville and Chattanooga Railroad, then on the Memphis and Charleston Railroad to Bridgeport Alabama. The bridge that crosses the Tennessee River there was burned by the rebels when Rosecrans advanced on Chattanooga and was currently being repaired. The wood had to come from Nashville, so it was taking some time.

October 23, 1863 The 102nd NY Volunteers regiment was sent to Nashville to convoy the Second Division train of wagons to Murfreesborough. General Grant approved opening a shorter and better route from Bridgeport to Chattanooga.

October 25, 1863 The Second Division was in Bridgeport, Alabama, getting ready for further action. They were on the banks of the Tennessee River and guarding the railroad from raids of rebel calvary. The First Division under General Williams was in Nashville guarding the railroad there. General Geary in his letter to his wife praised his son Edward who was an officer and captain of Battery F in Knap's Artillery. General Williams spent some time with the younger Geary and seemed to like him. The 102nd NY Volunteers were tasked with escorting the Division ambulance train from Murfreesborough to Bridgeport.

Reopening Tennessee River October 26–29, 1863

General Grant found a better route to get supplies to Chattanooga via Bridgeport and then defeated Bragg and drove him from Chattanooga which then opened the door to the Atlanta campaign in 1864.

The chief engineer of the Army of the Cumberland, Brig. Gen. William F. "Baldy" Smith, had devised a plan with Union General Rosecrans to open a better supply line to the troops in Chattanooga. General Smith briefed Grant immediately after the new commander's arrival and Grant enthusiastically agreed with the plan. Brown's Ferry crossed the Tennessee River from Moccasin Point where the road then followed a gap through the foothills, turned south through Lookout Valley to Wauhatchie Station, and then west to Kelley's Ferry, a navigable point on the Tennessee that could be reached by Union supply boats. If the Army of the Cumberland could seize Brown's Ferry and link up with Hooker's force arriving from Bridgeport, Alabama, through Lookout Valley, a reliable, efficient supply line—soon to become known as the "Cracker Line"—would be open. In addition, a force at Brown's Ferry would threaten the right flank of any Confederate movement into the valley. Excerpts taken from Cloudfront.net: Chattanooga Campaign

October 26, 1863 The Union Army 12th Corps Second Division, known as the White Star Division because of its white star on a blue background flag, was stationed along the Nashville and Chattanooga Railroad, as part of Union General Joseph Hooker's attempt to open the "cracker line" to bring supplies to Chattanooga, which was under siege. They drove off token rebel forces at Bridgeport, and then General Hooker split his command in two on the march through Lookout Valley, leaving the Second Division under General Geary at Wauhatchie Station, he took two divisions north three miles to guard the road to Kelly's Ford in order to protect the line of communications to the southwest as well as the road west to Kelley's Ferry.

October 28, 1863 Four Brigades of Hood's Division, 5000 men, surprise attacked the Second Division with its 1200 men, by coming from the east and the north. Picket fire began around 10 p.m. and the attack was in full mode by midnight. General Geary placed his regiments into a V shape to repel the rebels who came hard at them for over three hours. Geary's men continued to hold fast, though they began to run low on ammunition. Just as Confederate General Bratton began to sense victory, he received a note to retreat since Union reinforcements were arriving at his rear. The Second Division, which was not at full strength, successfully defended its position with the aid from the darkness. Confederate General Bratton withdrew to Lookout Mountain, successfully covered by Benning's Brigade. In the Wauhatchie fight, Bratton lost 356 men, while Geary's casualties numbered 216. However, General Geary's son Eddie was killed in this fight at Wauhatchie by sharpshooters of Hood's Confederate Division, who were instructed to fire at Union men who were highlighted by cannon muzzle flashes. The 102nd NY Volunteers were not involved in the action at Wauhatchie Station as they were busy convoying and guarding trains. General George S. Greene who commanded the Third Brigade was wounded and Colonel David Ireland from the 137th NY Regiment took temporary command.

The important part is that the Union Army now had its path to the outside and could receive supplies, weapons, ammunition, and reinforcements via the Cracker Line. The way was clear for the start of the Battles for Chattanooga on November 23.

October 31, 1863 The Third Brigade of the Second Division, which included the 102nd NY Volunteers, moved to Raccoon Mountain and took position on the hills to the right of the Second Brigade and started building breastworks. Union Gen. Ulysses S. Grant recognized Raccoon Mountain as an essential place in the campaign to resupply Federal troops in Chattanooga following the Battle of Chickamauga. Confederate forces had cut all supply lines into Chattanooga, and rations and materiel were running low. Grant realized that the mountain was lightly defended. A Confederate brigade managed the eastern slopes, and the 28th Alabama Infantry defended the approach from the Tennessee River.

November 2, 1863 The Second Division is camped between Raccoon Mountain and Lookout Mountain, about 3.5 miles from Chattanooga. They are in range of rebel artillery and had been shelled frequently. They had been building breast works and the rebel's shells were annoying but not very damaging. General Geary wrote a long letter to his wife about the battle at Wauhatchie and the loss of their son. He wrote, "The object of the movement upon which we have been engaged was to open a line of communication by which the main army under Thomas (late Rosecrans) was very nearly starved out and if some plan could be devised to get subsistence to Chattanooga, the army must fall back, the consequences of which no one could foresee, but that it would be exceedingly disastrous is admitted by all. The task devolved upon us. The Army is now fully supplied with subsistence and forage, and the men who so nobly periled their lives to save the Army, are every where cheered for various deeds. And we are hailed as the saviors of the Army of the Cumberland." APGTW p. 132 General Geary, the consummate politician was always writing about the greatness of his men and his command.

November 6, 1863 The Second Division of the 12th Corps is still camped near Wauhatchie, Tennessee. The official reports make it difficult to tell, but it seems like the 102nd NY Volunteers had rejoined the Second Division and were helping guard the railroad. The weather was warm, like an Indian Summer but also with lots of rain. There was a slight stalemate as the Union waited for General Sherman to arrive and Generals Thomas and Smith were tasked with coming up with a plan to attack Bragg's Confederates on Lookout Mountain and Missionary Ridge. The troops were busy fortifying their position and doing reconnaissance.

November 8, 1863 The Second Division of the Union Army 12th Corps is still in Wauhatchie fortifying their position. The weather is mild, but the stalemate continues. Rations for the men were still low and sometimes nonexistent for the animals. Even though the "Cracker Line" is open, the Confederates had killed more than 1,000 mules and horses of the Union Army and there were repercussions from that. General Geary wrote his wife, "The situation of the Army of the Cumberland has been, and still is, very precarious for want of subsistence and forage. The men being upon half rations, and many of the horses are without anything." APGTW p. 136

November 10, 1863 The Second Division is still on guard duty near Wauhatchie, the rebels up on Lookout Mountain are still shelling them with little effect. General Geary wrote his wife, "The booming of [Confederate] cannon from Lookout Mtn commences with the rising of the sun and continueth until the going down thereof, almost without intermission, yet we still have been so mercifully preserved as not to have received a single casualty. Occasionally our guns open fire upon the enemy for a few minutes simply for the purpose of shewing the enemy that our army still lives." APGTW p. 137

November 13, 1863 Moses Whitbeck who is now a Corporal in the 102nd NY Volunteers, is celebrating the second anniversary of his enlistment. He had one more year to go as he signed a three year enlistment. Also, General Sherman reached Bridgeport with the advance part of his troops. Soon after Generals Grant and Sherman went and inspected the fortifications the men had built facing the Confederate lines.

There had been some occasional firing at each other but no major skirmishes as each army was low on ammunition as well as rations as both sides were having supply issues.

November 14, 1863 The Second Division is still in Wauhatchie. By now they have strongly entrenched the area and are still waiting for the arrival of General Sherman. General Geary wrote his wife, "Our position remains unchanged with reference to the enemy. I am strongly entrenching my command, and I think can defend myself against largely superior numbers. Maj Gen [William Tecumseh] Sherman's command numbering 30,000 men are expected here in 3 days. What movement will take place I am unable to state." APGTW p. 138

November 17, 1863 General Sherman arrived in Chattanooga with his command. On paper it was 30,000 in the 15th Corps, but in reality, it was more like 23,000 were present for duty. In late October, Sherman had been promoted to Commander of the Army of the Tennessee. Now the generals will convene to determine which way to proceed, but the plan was that Sherman's arriving troops would use newly improved roads to pass through the hills north of Chattanooga, taking a route that was not visible from Lookout Mountain, hoping that Confederate Commander Bragg would not know for certain whether Sherman was targeting Chattanooga or Knoxville. Union General Smith would assemble every available boat and pontoon to allow Sherman's corps to cross the Tennessee River near the mouth of the South Chickamauga Creek and attack Bragg's right flank on Missionary Ridge. If the attack were successful, the Union would control the two key railroad lines that supplied Bragg's army, forcing him to withdraw. Thomas's army, which included the 102 NY Volunteers, would simultaneously pin down the Confederate center on Missionary Ridge. The plan also called for Hooker to assault and seize Lookout Mountain, Bragg's left flank, and continue on to Rossville, where he would be positioned to cut off a potential Confederate retreat to the south. Excerpts taken from Wikipedia: Chattanooga campaign

Although Grant had hoped to begin offensive operations on **November 21**, by **November 20** only one of Sherman's brigades had crossed over Brown's Ferry and the attack had to be postponed. Grant was coming under pressure from Washington to react to Confederate General Longstreet's advance against Union General Burnside at Knoxville. Confederate General Bragg worsened the rebel's situation on **November 22** by ordering Maj. Gen. Patrick R. Cleburne to remove his and Simon B. Buckner's divisions from the line and march to Chickamauga Station, for railroad transport to Knoxville, removing 11,000 more men from the defense of Chattanooga. This move was apparently made because, as Grant had hoped, Bragg concluded that Sherman's troops were moving on to Knoxville, in which case Longstreet would need the reinforcements, for which he had been constantly clamoring since he was first given the assignment of attacking Knoxville. Excerpts taken from Wikipedia: Chattanooga campaign

Chattanooga-Ringgold Campaign November 23–27, 1863 The 102nd NY Volunteers were very engaged during this action. Luckily, Colonel Lane who commanded the regiment wrote a detailed report, something that had been lacking in the War of the Rebellion official record until now.

Battle of Lookout Mountain November 23–24, 1863 The 102nd NY Volunteer Regiment, as part of the Third Brigade of the Union Army 12th Corps Second Division under

General John W. Geary, took a significant role in this battle. General Geary wrote a long report that at times waxed poetic, while Colonel Lane's report was the facts only. Colonel Ireland commanding the Third Brigade also wrote a report. All of these reports are available in the War of the Rebellion Series.

November 23, 1863 Union forces under General Sherman captured Orchard Knob. They had planned just a reconnaissance but because the 14,000 strong Union force quickly overtook the 600 rebels they decided to stay. Luckily for the Union, the confederates split their forces that day between the shelf at the base of the mountain and the mountain top itself. But, they made a critical error in putting the troops at the top at the actual ridge instead of just below it where they had a better view and defense of the ground below, they also withdrew Maj. Gen. William H. T. Walker's Division from the base of Lookout Mountain and placed them on the far right of Missionary Ridge, just south of Tunnel Hill. This movement would make a difference the next day. Excerpts taken from Wikipedia: Chattanooga campaign

General Geary wrote his wife, "The scenes around us are anything but peaceful. My troops have been in line of battle for two days. The roar of artillery & the rattle of musketry do not cease from morning's dawn to the latest gleam of evening. We have just been reinforced by Gen Sherman's Corps from the Mississippi, numbering 20,000 men." APGTW p. 141

November 24, 1863 The 102nd NY Volunteers left their camp at 6:15 a.m. and traveled with the Third Brigade to the foot of Lookout Mountain about 8 a.m., where they found the Second Brigade "resting on their arms." That meant that they had bivouacked with their weapons. After about a half hours' rest, the whole command, some 4,000 men, climbed the mountain by a small trail near a white house, about a mile and a half south of the dividing ridge, between the north and west sides of the mountain. The 102nd NY Volunteer regiment sent forth skirmishers using about half of their men, the rest held in reserve. After about a ¾ mile push the regiment was halted so the rest of the brigade could catch up. Once the brigade caught up, the skirmishing continued with a strong rebel force on the Union left.

Colonel Lane reported, "On coming to the first rebel camp we met with a very spirited resistance, and I detached two companies (one from the right and one from the left) of the reserve to strengthen the skirmishers and to flank the rebels. In three minutes thereafter we had possession of the camp and took about 50 prisoners. These were sent to the rear under guard, as were numerous others, and I supplied the gaps made in the line of skirmishers (by their taking the prisoners off) from the reserve. We then pushed on, taking another camp, and led the brigade to the backbone dividing the north and west sides of the mountain. Here I halted the reserve, by orders received from an aide to Gen. Hooker not to pass the ridge, and gathered all the skirmishers possible, closing them on the reserve. The Third Brigade went by on the double-quick, and Capt. Stegman and some 10 or 12 skirmishers with them, driving the rebels some distance past the white house, on the north side of the hill. I halted my reserve about 200 feet from the perpendicular rock at the crest. While lying here, the western troops passed a little way over the backbone, some regiments going as far as the white [Craven] house. Just at this time the rebel sharpshooters opened fire on my reserve from the perpendicular rocks over our heads, and not being in a position to return it, I moved the regiment down the hill to the

partly completed battery on the brigade. This regiment numbered 130 muskets on entering the fight. We took over 620 prisoners, and took them safely to the rear. The regiment lost in this engagement Maj. G. M. Elliott, killed (a gallant officer), Lieut. Col. Robert Avery, thigh broken by a Minnie ball (since amputated), First Sergt. R. Mulholland, Company H, wounded, Private David Hunter, Company A, wounded."
WOTR Vol 31 p. 444

At the Cravens House, the Mississippi brigade of Gen. Edward Walthall tried to resist the Union tide but failed. While Col. David Ireland's New York brigade pressed Walthall's front, Col. George Cobham's Pennsylvanians were delivering a withering flanking fire from the upper portion of the slope.

Colonel Ireland from the 137th NY Volunteer Regiment was leading the Third Brigade and reported, "We had been engaged with the enemy since 10 a.m., had marched over great natural obstructions for over 4 ½ miles, fighting the enemy at every step; had driven them from every position; taken prisoners all the forces that were on the mountain, among them a large number of field officers; captured what the rebels termed the Gibraltar of America, and held it until relieved."
WOTR Vol 31 p. 437

The Union force was just too strong, and the confederates were weak and stretched too thin. The Union was also helped by the fog which made their movements difficult for the rebels to see. By 2 p.m., the "Battle Among the Clouds" was over. The Confederate command had mismanaged their defense and overestimated the advantages offered by the mountain, and 1,200 rebels faced nearly 12,000 attacking Yankees. Confederate artillery proved of little use, as the hill was so steep that the attackers could not even be seen until they appeared near the summit. The troops had been placed at the wrong place at the top of the hill. General Bragg did not send reinforcements because the Union attack against the Confederate center (involving the 102nd NY Volunteers) was more threatening than the sideshow around Lookout Mountain.

The Confederates abandoned the mountain by late afternoon. That night Bragg held a counsel with his generals and decided to withdraw from Lookout Mountain to reinforce Missionary Ridge, The next day, Union forces launched a devastating attack against Missionary Ridge giving Grant the second victory of the fight for Chattanooga and successfully breaking the Confederate lines around Chattanooga.

General Geary wrote in his report a tribute to Major Elliott of the 102nd NY Volunteers who was killed. He said, "Maj G.M. Elliott, entered the service with his regiment as first lieutenant, in October 1861. In the fall of 1862, he was detailed as ordnance officer of this division, and was relieved in July of the present year to enable him to accept the rank he held at the time of his death. During his entire connection with this command he exhibited qualities and combinations of character of the rarest development; thoroughly educated as a scholar and a gentleman, he was no less an accomplished soldier and a strict disciplinarian.

"Possessing a remarkable degree the impetuosity of the young soldier, with the cool and cautious prudence of the veteran, had he lived the profession of arms could not but have been adorned by his association. When he fell, the first one shot in the division on Lookout Mountain, the command lost one of its brightest ornaments; one cherished, respected and esteemed, by all grades, as possessing the noblest character- istics of each." WOTR Vol 31. Part II p. 407

Battle of Missionary Ridge November 25, 1863

November 25, 1863 In the morning, elements of the Union Army of the Tennessee commanded by Maj. Gen. William Tecumseh Sherman attempted to capture the northern end of Missionary Ridge, Tunnel Hill, but were stopped by fierce resistance from the Confederate divisions of Maj. Gen. Patrick Cleburne, William H.T. Walker, and Carter L. Stevenson. In the afternoon, Grant was concerned that Bragg was reinforcing his right flank at Sherman's expense. He ordered the Army of the Cumberland, commanded by Maj. Gen. George Henry Thomas, to move forward and seize the Confederate line of rifle pits on the valley floor, and stop there to await further orders. The Union soldiers moved forward and quickly pushed the Confederates from the first line of rifle pits, but were then subjected to a punishing fire from the Confederate lines up the ridge.

At this point, the Union soldiers continued the attack against the remaining lines, seeking refuge near the crest of the ridge (the top line of rifle pits was situated on the actual crest rather than the military crest of the ridge, leaving blind spots). This second advance was taken up by the commanders on the spot, but also by some of the soldiers who, on their own, sought shelter from the fire further up the slope. The Union advance was disorganized, but effective, finally overwhelming and scattering what ought to have been, as General Grant himself believed, an impregnable Confederate line. In combination with an advance from the southern end of the ridge by divisions under Maj. Gen. Joseph Hooker, the Union Army routed Bragg's army, which retreated to Dalton, Georgia, ending the siege of Union forces in Chattanooga, Tennessee. Excerpts taken from Wikipedia: Battle of Missionary Ridge

In the morning, the 12th Corps Third Brigade buried their dead. Then the Division marched just before noon for Missionary Ridge. Arriving late in the day, the Second Division reached the top around 6 p.m.

Colonel Ireland commanding the Third Brigade that had the 102nd NY Volunteers, reported, "After crossing Chattanooga Creek I was ordered to remain there until the artillery had crossed, and then guard it. In accordance with these orders I moved from the creek to Rossville Gap in the rear of the artillery. At the gap, the command turned to the left along the base of the ridge. After marching about a mile from the gap I received orders to move the brigade forward in column of regiments to the support of the troops on the ridge. After moving about 2 miles in this manner, always within sound of the musketry of our advance, but without being engaged, we bivouacked for the night in an old rebel camp at the base of Missionary Ridge." WOTR Vol 31 Part 11 p. 437

Colonel Lane reported about the 102nd NY Volunteers, "this brigade as escort and guard to the artillery. This regiment, in common with the brigade, acted as support to the batteries, which, by shelling the ridge, terminated that fight." WOTR Vol 31 Part II p. 444

November 26, 1863 The 102nd NY Volunteers marched toward the enemy, but they did not make a defense. The Second Division however skirmished with the rear guard of Confederate General Breckinridge's Corps at Pea Vine Creek and drove them away. The 102nd camped on the other side of Chickamauga Swamp near Pea Vine Creek, but they didn't really camp as they had no blankets, but General Geary reported that the men were cheerful and eager to finish the task at hand.

Battle of Ringgold Gap, Taylor's Ridge, November 27, 1863 The Third Brigade left Pea Vine Creek at 7 a.m. and arrived at Ringgold about 10 a.m. After they crossed a covered bridge at Catoosa Creek, they came under fire from musketry and artillery from Confederate General Cleburne's Division. They hid behind the depot there and the railroad embankments until it was time to move.

Colonel Lane reported, "started for the enemy, and at about 11 a.m. found the enemy had possession of the range of hills beyond Ringgold. The brigade was marched through the town, and halted before the stone depot, being held in reserve. After having been here about an hour, it was reported that the western troops were falling back from the gap, on the right, and this regiment, with the brigade, was sent some three-fourths of a mile, at a double-quick, to relieve them; we marched through a heavy fire of musketry and artillery to the desired point, three charges of grape and canister going through our (the brigade's) ranks while we were passing the wet swamp land. Capt. Greene, assistant adjutant-general, Third Brigade, lost his leg from one of the artillery discharges. At this time the regiment was gotten into position, as quickly as possible, but the men of regiment were much mixed; those fleetest and most enduring being to the front and left, and the others in a line along the creek. Lieut.-Col. Randall, One hundred and forty-ninth New York, took possession of a small barn on the left, and, assisted by men of the several regiments, held the enemy at bay. The men, although thoroughly exposed to the aim of the rebels, and not allowed to return the fire, were cool, never offered to flinch or retreat, and when any were wounded they obeyed their officers and remained quiet (many not asking to be taken to the rear) until the firing ceased. The conduct of the men was more than good; it was heroic. There was much danger at this time that the brigade would be flanked, and word was sent to that effect to the rear; but although the balls came thick and fast, the men stood firm, and when Knap's battery opened on the rebels, sending the shells over our heads into the rebels, it was difficult to keep the men from rising and cheering. Very soon after our artillery opened, at this time, the rebels' firing slackened, and soon ceased entirely. At this time Capt. Stegman, assisted by Lieut. Davies, both of this regiment, with 30 men of the One hundred and second and 10 men from the Sixtieth New York Volunteers, went (by orders) through the pass, using half his men as skirmishers, and by a few volleys cleared the pass completely, and arrived in sight of the railroad bridge beyond in time to fire on and disperse the rebels, who were firing the bridge to check the pursuit. The reserve put out the fire while the skirmishers pushed on, driving the rebels form the bridge beyond, some eighth of a mile farther." WOTR Vol 31 Part II p. 445

The truth is that at Ringgold, Confederate Major General Cleburne's men fought stubbornly to protect the army, which they did by compelling Geary's men to stand and fight. The Confederates successfully safeguarded the retreat and punished the pursuers. Because of Cleburne's tough defense of Ringgold Gap, the retreating Confederate army survived to fight on another nine months in the North Georgia hills.

November 28, 1863 The Third Brigade under Colonel Ireland, which included the 102nd NY Volunteers and Corporal Moses Whitbeck, was put on picket duty along the ridge on both sides of Ringgold gap. The approaches were strengthened by slashings, detachments were scattered, and many large fires built to give the rebels the impression that a large force was there.

November 29, 1863 The Third Brigade of the Second Division of the Union Army 12th Corps was relieved by the First Brigade and went to the houses in Ringgold for shelter. Large details of men were sent to destroy the railroad the next few days.

December 1, 1863 The Second Division destroyed all the bridges and left fires burning on the mountain and then marched to their old camp in Lookout Valley. The campaign had cost them 138 casualties at Lookout Mountain and another 203 at Ringgold. This would be their last major action of 1863.

December 12, 1863 The Second Division is camped near Wauhatchie, Tennessee. The rainy season had begun and ended the campaigning by both armies. General Geary wrote his wife, "The rainy season is just setting in here, and rains as if Noah's flood was about to be re-enacted. The roads are muddy, the streams overflowing their banks, and our campaign seems to have a veto placed upon its further progress by the hand of nature. But we think we have marched enough, and fought enough to entitle us to a little rest. If man will not give it, nature will." APGTW p. 145

The whole Division turned out to give the 29th Pennsylvania Volunteer Regiment a sendoff for their 30-day furlough for re-enlisting and to go recruit more members. The War Department was worried about the large amount of three-year enlistments that would end in the summer and fall of 1864, so they offered them furloughs in the winter if the whole regiment re-enlisted then. They would also receive the designation as Veteran Volunteer units.

December 19, 1863 The Second Division is still camped near Wauhatchie, Tennessee. Many regiments in the command are re-enlisting and the commanders are busy taking care of all that entails.

December 31, 1863 As 1863 came to a close, the weather in Wauhatchie Tennessee where the 102nd NY Volunteers were stationed became extremely cold. General Geary wrote his wife, "the weather set in extremely cold…, and the ground has been frozen several inches in depth ever since. The mountains are white with snow. The whole idea of "Sunny South" is exploded, and much of the poetic idea of Southern beauty is with us fully exploded. It is a humbug and a false-hood, and a lie which this war has fully exploded. Southern greatness was always a humbug in my opinion, and more so now than ever before." p. 150

From research it appears that the winter of 1863-64 was a hard winter and would lead to late operations beginning in the spring of 1864. Different diaries tell about a hard freeze around **January 1** lasting until **January 10**, and a cold winter overall. This would preclude any military maneuvers.

1864

January 1, 1864 Moses Whitbeck was promoted to sergeant. The 102nd NY Volunteers were given a 30-day furlough that started on this date (WOTR Vol 31 p. 27) as an enticement to re-enlist.

January 30, 1864 Moses Whitbeck re-enlisted as a veteran. The men of the 102nd New York Volunteers were now called Veteran Volunteers to distinguish them from new units and also from the Federal army that existed before the war, and also as an enticement to re-enlist. The 102nd NY Veteran Volunteers were now part of the Army of the Cumberland in the 20th Corps, Third Brigade in the Second Division. As part of their re-enlistment, the men of the 102nd received a 30-day fur-lough. Now the Third Brigade is in Stevenson while the rest of the Second Division is in Bridgeport, Alabama. We know from his book that General Geary was absent on furlough from mid-January until mid-February. It is very likely that the corps commanders received time off as well as their divisions. APGTW p. 150-151

The records indicate that in the spring of 1864, the 102nd NY was on duty in Lookout Valley until May 1864. Lookout Valley was a strategic point in Tennessee near Chattanooga and was winter quarters for the Army of the Cumberland. Chattanooga was called the "Gateway to the Lower South," and the Union Army was not going to give it up now that it had cost them dearly to possess. General Geary lists on his letters that the Second Division HQ is in Bridgeport, Alabama. The WOTR reports tell us that the 102nd NY Volunteers were with the Third Brigade in Stevenson Alabama.

In **February 1864**, The Third Brigade of the Second Division was in Stevenson Alabama guarding the railroad and performing provost duty (which is basically Military Police and sometimes includes guarding prisoners). On **February 11**, four companies of the Third Brigade were sent to Anderson, Tennessee to guard that post.

February 20, 1864 The Second Division received information that Confederate General John Morgan intended on making a raid with the aim of destroying the railroad bridge over the Tennessee River and tearing up the track of the railroad in that area. Every effort was made to prepare for the attack and then on **February 21** the 147th Pennsylvania was sent on a reconnaissance to find the rebels but did not find them.

February 22, 1864 General Hooker received a telegram stating that Cleburne's Confederate Division was at La Fayette, Georgia and intended to attack the railroad opposite Bridgeport or between there and Chattanooga. Once again, the Division prepared for the attack, but it never came. They now had a line of pickets that covered three miles above Bridgeport, to a point 1 ½ miles south of Anderson, Tennessee. They were guarding the fords on the river and patrolling the railroad.

February 27, 1864 The Third Brigade is still in Stevenson, Alabama. The men are busy looking for and preparing for rebel raids and General John Morgan. General Geary wrote his wife Mary, "I am exceedingly busy watching for the raids with which this place is constantly threatened by John Morgan and his ilk, for if they do come they will be received with the highest military honors, and with a warmth he has seldom or perhaps never felt." APGTW p. 154 Some skirmishing had taken place south of Ringgold and every preparation was being made for the spring offensive.

March 1, 1864 The Third Brigade was still in Stevenson Alabama guarding the railroad. The War of the Rebellion series here reports that the 102nd NY Volunteers returned from their furlough on **March 12th**. WOTR Vol 32 p. 29 General Geary wrote, "Coming events cast their shadows all around us in the Southwest, and if we can judge aright from the movements of our Armies, some blow must soon be struck which if successful, (and I have no doubt it will be so) will break the backbone of rebeldom. God grant it may be so and that the benign influencers of peace may soon and forever extend through out the land." APGTW p. 155

March 4, 1864 The Second Division of the Union Army 20th Corps under General John W. Geary will soon be on the march but for now are still in Bridgeport and Stevenson. The climate is mild, and Spring is near. The Tennessee River is almost overflowing its banks, and it is raining. APGTW p. 156

March 18, 1864 The Division still had not marched, but General Geary had taken some troops from two regiments from the First Brigade on a reconnaissance toward Trenton and returned this date.

General Geary wrote, "I have just returned from an important reconnaissance in Georgia, in the direction of Trenton and South of that place, which is about 16 miles South East of this place. I was out two days, bivouacking on the ground at night.

"The country through which I travelled with 1000 men, without wagons, ambulances, or artillery was almost an unbroken forest for 14 miles. I dispersed some small bands of the enemy and captured 6 prisoners without any loss on my side. I gained much important information concerning the country, the position and condition of the enemy." APGTW p. 159

The 102nd NY Veteran Volunteers were part of the Third Brigade and so not involved.

March 21, 1864 The Second Division is still in Bridgeport, Alabama. General Geary wrote his wife, "The enemy continues to move about as if he sometimes has the intention of attacking our lines, but seems to be undecided as to where, how, or when. The busy hum of preparation is heard on all sides. The continued movements of troops indicate that sooner or later their marches must bring men into collision with the enemy, and a battle be fought. God grant that we may be victorious and the war soon terminated. If you could only see the horrors of war for a single day you would never desire to look upon its like again, and would not wonder why your soldier husband hates it so much, yet duty bids us stay and endeavor to bequeath peace at last to our children for their heritage. I often feel a desire to be translated back to the days of my romance, and enjoy the pleasure of air castles and gorgeous imagery of youth, but the facts, stern and frightful, that are continuously surrounding us makes me feel that all is real and that true happiness mocks us as we follow it like will-o-the-wisp, the puerile grasp at the ever fleeing rainbow." APGTW p. 160 As for the Third Brigade that contained the 102nd NY Volunteers, they were still camped near Stevenson, Alabama.

On **April 4, 1864**, just before the onset of the Atlanta Campaign, William T. Sherman authorized the consolidation of 11th and 12th Corps as the 20th Corps, under Hooker's command, to serve in the Army of the Cumberland. The Army of the Cumberland as well as the Army of Ohio and the Army of Tennessee were all commanded by Union General Sherman. General Grant was now the General-in-Chief.

Because of the new corps, General von Steinwehr of the 11th Corps resigned and General Slocum of the 12th Corps accepted a transfer to a smaller command on the Mississippi.

General Geary continued to command the Second Division and now reported to General Hooker who he had a good relationship with. The transition bothered some of the soldiers. The "White Stars" were not on good terms with the "Teutonic crescents" from the days of Chancellorsville and also "Wauhatchie" where they did not come up in time to assist in repulsing Longstreet's superior numbers.

April 9, 1864 The 20th Corps Second Division was still camped near Bridgeport, while the Third Brigade that included the 102nd NY Veteran Volunteers and Sgt. Moses Whitbeck were still in nearby Stevenson. The roads that spring were still bad, as the weather was cold and wet, and General Geary guessed that the corps would not move until somewhere near May 1. General Geary wrote his wife, "I have just returned from a trip over the late battle-fields of Lookout, Mission Ridge and Ringgold. The tour occupied three days, and I assure you was very instructive and interesting. The weather still continues cool and wet, and of course, the roads are in such a condition as to preclude any idea of a movement or active hostilities until near the 1st of May, and which time I think it is probable that not only this but all the other armies will move simultaneously, and the grand crisis of the war will be upon us. We cannot tell of, or know how it will result, until after the battles are fought, but I have an abiding faith, with God's favor, the campaign will be decisive." APGTW p. 163

April 11, 1864 The 102nd NY Volunteers were involved with a reconnaissance to Caperton's Ferry, Alabama to arrest several prominent citizens residing on the south bank of the Tennessee River, near Caperton's Ferry. Captain Stegman reported, "The detachment commenced its march at 4 a.m. on the morning of the 11th, and proceeded directly to the north bank of the river. There, with the aid and assistance of Lieut.'s Merriam and Brown, One hundred and forty-ninth New York Volunteers, we were quickly embarked in scow, dug-outs, and pontoon-boats, and, after much difficulty, succeed in effecting a crossing. Immediately upon reaching the south bank I deployed a strong line of skirmishers, under command of Lieut. Kelsey, and marched swiftly up the road to our first point of destination. In the mean time, however, I had captured two young men lounging near the river bank, and impressing one to act as a guide I forwarded the other to the north bank under guard, to be held as hostage for the good behavior of his brother. We reached the residence of Mr. Hugh Caperton, and, discovering said person in an adjacent field, I immediately arrested him. Following the lower mountain road, under the direction of our guide, I filled to the right, halting for a movement at the house of a Mr. Marshall, a citizens desirous of taking the oath, and, after some conversation, gaining information, I proceeded onward, arresting Mr. Adam Caperton, and discovering by search and inquiry that Mr. Thomas Caperton, one of the parties noticed for arrest, was a soldier in the rebel service, and had not been at home or seen in his immediately neighborhood for several months past. Retracing our steps, throwing out another line of skirmishers to our then front and holding our former first line as rear guard, I advanced to the left of Mr. Hugh Caperton's and advanced to the residence of Mr. John E. Caperton. This person I discovered to be absent from home, having gone to the top of the mountain. From searching inquiry I became convinced that this man has been endeavoring for more than a week to reach Stevenson for the purpose of taking the oath of allegiance. We then proceeded to the late residence of Mr.

Sam. Norwood, finding, however, that he had long since vacated, removing to some inner county, his present place of residence. I arrested the man who at present occupies the premises first named, a person named John Loweree. The house noticed on the map as Norwood's house, near the coal bank, on the mountain top, has been utterly destroyed by fire. In each case I made thorough investigation, searching the premises for all articles contraband of war, but discovered nothing. Houses and outhouses, pens for animals, everything bearing the look of a depository for guns or Government property were diligently scrutinized, without effect. Having accomplished the object of my mission, to the extent of my ability, and believing that further search would be as ineffectual and fruitless as previous search has proven, I returned to the ferry and north bank of the river, thence to Stevenson, where I delivered the persons of Messrs. Hugh and Adam Caperton and Mr. John Loweree into your charge and keeping.

"In the course of my investigations I became acquainted with the fact that a strong guerrilla rendezvous exists at Raccoon Creek, about 7 miles from Caperton's Ferry, under the leadership of a person named Cox. This man, with some 15 or 20 comrades, has dashed through the valley on Thursday or Friday last, committing serious depredations. This man is the same persons who attacked the detachment of the Sixty-sixth Ohio a few weeks ago. A rumor prevails in the valley that some 1,500 of Morgan's men are congregated in the mountains. Numerous individuals in the valley and on the mountains are desirous of taking the oath of allegiance.

"The arrested Caperton brothers are considered the wealthiest and most influential men in the valley. Both have nephews and sons in the rebel army; Mr. Loweree has two sons in the service of the rebels.

"The roads are in execrable condition, miry and rocky.

"I must render proper thanks to the officers and men assisting me, all from the Sixtieth New York, for alacrity and obedience to every command and the endeavor to do more than I required. I cannot too highly compliment them.

"The expedition is under many obligations to Lieut. Merriam and Brown, of the One hundred and forty-ninth New York, for their diligence and attention in ferrying the command over the river and return." WOTR Vol 32 p. 656-657

April 12, 1864 The Second Division under General Geary took some troops and went on a reconnaissance down the Tennessee River on the steamboat Chickamauga to clear the river, but the 102nd NY Veteran Volunteers were not involved as they had been engaged the day before. The force returned on April 15, having destroyed 47 boats and capturing four prisoners and gaining important information about where the rebels were. WOTR Vol 32 p. 33

April 14, 1864 The organization of the Union Army 20th Corps was made official this date by Special Field Order No. 105, Department of the Cumberland. The Second Division which included the 102nd NY Veteran Volunteers, received the addition of the 11th Corps Second Division First Brigade, which included the 119th NY Volunteers. This addition brought the Division up to a strength of 20 Regiments and 12 pieces of artillery.

May 2, 1864 The Second Division of the 20th Corps received their marching orders and prepared to move out. The 102 NY Volunteers broke their camp at Stevenson, Alabama, to start the Atlanta Campaign. They were part of the Third Brigade and had to march over to Bridgeport to join the rest of the Division.

Fort Harker, located near Stevenson, Alabama, was a military fortification built by the Union Army in the summer of 1862 by soldiers and freed slaves of the Army of the Cumberland. The fort helped secure strategic railroad lines to ensure the free movement of Union troops and supplies in southeastern Tennessee and northeastern Alabama. Union General William Rosecrans established his headquarters at Fort Harker in July, 1863, from where he directed a successful campaign against the position of Confederate General Braxton Bragg in Chattanooga, Tennessee.

Although the Corps was consolidated and ready to move on the afternoon of May 2, the orders came to wait until the next day to actually start the campaign.

On **May 3, 1864**, the Second Division marched to and reached Lookout Valley, Tennessee, southwest of Chattanooga. Lookout Mountain had been the sight of the battle in **September 1863** where a Federal army led by General William S. Rosecrans was besieged there by a Southern army commanded by General Braxton Bragg. In **October 1863**, General Ulysses S. Grant took over the campaign to relieve the Union troops and seize the offensive. With the help of reinforcements from General Joseph Hooker and General William Tecumseh Sherman, including the 102nd NY, the Federal forces defeated the Confederates in the Battles of Lookout Mountain and Missionary Ridge and lifted the siege; by the end of the month, the Confederate army was in retreat into Georgia.

As for the Third Brigade under Colonel David Ireland, they marched at 9 a.m., crossed the Tennessee River, then marched to Shellmound so the Division could consolidate even more there as other commands joined back in the corps.

May 4, 1864 The Second Division marched early because the weather was humid. They passed through Whiteside's and Lookout Valley and crossed Lookout Mountain and then camped in the Chattanooga Valley having traveled 22 miles. The 102nd NY Volunteers went with General Kilpatrick on a reconnaissance to Snake Creek Gap.

By May 7, 1864, Maj. Gen. William T. Sherman's army group had crossed southward into Georgia and advanced toward Gen. Joseph E. Johnston's Confederate army defending Atlanta. General Johnston had entrenched his army on the long, steep Rocky Face Ridge between the Western & Atlantic Railroad and Dalton, and eastward across Crow Valley. When Sherman approached, he demonstrated against this position with two armies, the Army of the Cumberland and the Army of the Ohio, while he sent a third one, the Army of the Tennessee, through Snake Creek Gap, to the south, to hit the Western & Atlantic Railroad at Resaca, cutting Johnston's retreat and resupply route. The first two armies engaged the enemy beginning on May 8 at Buzzard Roost (Mill Creek Gap) and at Dug Gap while the third army, under Maj. Gen. James B. McPherson, passed through Snake Creek Gap and on May 9 advanced to the outskirts of Resaca, where it found Confederates entrenched. Fearing the strength of the enemy, McPherson pulled his column back to Snake Creek Gap. On May 10, Sherman decided to join McPherson in an effort to take Resaca. The next morning, Sherman's army withdrew from in front of Rocky Face Ridge. Discovering Sherman's movement, Johnston retired south toward Resaca on May 12.

The five-day-long battle along Rocky Face Ridge was the first battle of the Atlanta Campaign.

Though, the first battle of the Atlanta Campaign had been a victory, McPherson's march went undetected. Sherman's move to take Atlanta was well underway.

The bluffs of Rocky Face presented William Tecumseh Sherman with his first challenge in the Atlanta Campaign. Joseph E. Johnston had chosen the site and prepared well for the Federal onslaught. Steep cliffs give way to a high gap called Mill Creek, which locals refer to as "Buzzards Roost." Just over two miles further south is Dug Gap, an equally imposing fracture in the mountain that runs along the western edge of Dalton, Georgia. To the north of the city the mountain dwindles to the floor of Crow Valley. When Sherman inspected the field, he called it "doors of death" and gave the job of taking Rocky Face to General George Thomas. The Union commander tested the area after capturing Tunnel Hill on May 7. Excerpts taken from MyCivilWar.com: The Battle of Rocky Face Ridge

The next day, **May 8, 1864**, General Thomas ordered troops to attack Dug Gap where well-entrenched rebel forces used large boulders to turn back bluecoats that outnumbered them 10 to 1. Union troops of the 20th Corps, Second Division, under General John W. Geary advanced up the steep slopes at a place called Dug Gap. Geary's men, about 4500 strong, many of whom had scaled Lookout Mountain six months earlier, reached the Confederate line composed of two regiments of Arkansas infantry and the 9th Kentucky Calvary and was about 1000 men strong. Hand-to-hand fighting broke out along the craggy mountain crest as daring Yankees vaulted over rocks and boulders to break the rebel line. Ultimately, Confederate reinforcements under the command of General Patrick

Cleburne arrived on the scene and successfully drove the Federals back down the mountainside. Stung by the defeat of his larger force, Thomas decided to probe the line further north, near Buzzard's Roost, while General James McPherson was moving south to outflank the rebels in a pattern Sherman would repeat many times over the next five months. Excerpts taken from MyCivilWar.com: The Battle of Rocky Face Ridge

The Second Division, Third Brigade, including the 102nd NY Volunteers and Sgt. Moses Whitbeck, had been separated from the rest of Geary's division and were sent with their commander, Colonel David Ireland, to help Union forces under General McPherson at Snake Creek Gap where they successfully broke through Confederate lines. Colonel Lane Reported, "on Sunday, May 8, this regiment, as a part of the Third Brigade, left the division column, and as escort to cavalry under Gen. Kilpatrick made a reconnaissance of eight miles to form a connection between our column and that under Gen. McPherson. On arriving within half a mile of the place for junction the brigade was halted and formed line of battle and this regiment, with others, erected breast-works, after which the men rested until about 3 p.m. In the mean time the cavalry under Gen. Kilpatrick had succeeded in making connection with McPherson's corps, a portion of which command relieved us, when we rejoined our division by marching about eight miles and arrived at Dug Gap, in the Rocky Face Ridge, about one hour after dark, but not in time to participate in the attack made by the division during the afternoon. The brigade was put in close column of regiments and the men slept on their arms." WOTR VOL 38 pt. 2 p. 293

On **May 9, 1864**, The 102nd NY Volunteer Regiment was ordered back about 200 yards from where they had spent the night sleeping "on their arms." They spent the day building breastworks about the width of 1.5 times their front. They again spent the night on their arms and each company took turns standing guard that night. Thomas H. Elliot, the assistant adjutant-general to General Geary, wrote by command of General Geary, "The general commanding division takes pleasure in announcing his appreciation of the gallant conduct of the troops of his command in their assault upon Rocky Face Ridge yesterday, in execution of peremptory orders to attempt to take the gap leading through it. The troops of the division, by their exhibition of valor in assaulting the almost impregnable position of the enemy, sustained its proud prestige. Night approaching and the mountain offering no shelter for the troops, and our engagement of the enemy having diverted his attention from General McPherson's advance and enabled him to pass through Snake Creek Gap south of us, you were withdrawn to encamp. You have accomplished an object of great bearing upon the success of the present movements." WOTR pt. 2 p. 148

On the morning of **May 11, 1864,** Confederate General Carter Stevenson woke and sensed a problem. The gunfire of the past three days had stopped. He immediately communicated his suspicion to Gen. Johnston, who ordered Wheeler on a scouting mission west of Rocky Face. Wheeler confirmed that the entire Federal Army had withdrawn and was apparently heading south along the west side of Taylor Ridge to hide the movement. Faced with an opponent of superior numbers, Johnston had no choice but to withdraw south to Resaca. The battle of Rocky Face was over, not with a bang but a whimper. Union General Geary's Second Division, which included the 102nd NY Volunteers and Sgt. Moses Whitbeck, remained in place May 9-11 at the foot of the Rocky Face mountain and guarded the approaches to and from it for a distance of five miles. Excerpts taken from MyCivilWar.com: The Battle of Rocky Face Ridge

On **May 12**, the entire Second Division under General Geary marched to and through Snake Creek Gap, camping a short distance from its eastern opening.

May 13, 1864 The Union Army of the Cumberland's 20th Corps including the 102nd NY Volunteers and Sgt. Moses Whitbeck moved toward Resaca under arms and ready for a fight. The men had been carrying overcoats and sleeping out in the open without tents. The Georgia days had been hot, and the nights quite cold and foggy. Most of the soldiers and officers had been without a change of clothing for nearly three weeks. But the 20th Corps had no time to waste as the Battle of Resaca was upon them. The Regiment had marched in the daytime and that night they built breastworks in a place called Sugar Valley. The slept that night "on their Arms."

May 14, 1864 In the Battle of Resaca, Johnston's Confederate army was positioned north and west of Resaca, stretching four miles, with its left on the Oostanaula River and the right extending to the Connasauga River. Sherman's forces, marching from Snake Creek Gap paralleled the rebel lines. As they did, Sherman ordered attacks to keep the rebels occupied while Sweeny's division crossed the Oostanaula four miles downstream from Resaca at Lay's Ferry, beyond the Confederate left, to threaten the railroad.

The Union Army of the Cumberland 20th Corps moved through thickets and underbrush to the rear and in support of General Daniel Butterfield's 3rd Division, and in the afternoon received a hurried order to move quickly farther to the Union left to support one brigade of Union General David S. Stanley's Division of the 4th Corps. The rebels had broken Stanley's Division and were pressing it with yells and were nearing one artillery battery of the 5th Indiana when the 20th Corps Third Brigade, came from the woods across where the 5th Indiana was and surprised the rebel force. The rebels fled without much resistance. General Alpheus Williams wrote, "They skedaddled as fast as they had advanced, hardly exchanging a half-dozen volleys. They were so surprised that they fled wildly and didn't wound a dozen men." FTCM p. 308 The 102nd NY was tasked with helping the 111th Pennsylvania regiment to hold the extreme front of the Union line.

Meanwhile on the morning of **May 14**, Federals of Maj. Gens. John Schofield's III Corps and Oliver O. Howard's IV Corps attacked across Camp Creek but failed to dislodge their opponents. Stiff resistance by Confederate Major Generals Patrick Cleburne's and Thomas Hindman's divisions helped repel the attacks, with Maj. Gen. William Bate's division bearing the brunt of the fighting. The Federal assault on the Confederate center-right petered out around 3 p.m. The heavy fighting of the day soon slowed to desultory picket fire. Still, Johnston thought he controlled the field and decided to send Hood against the Federal left the next morning.

On the morning of **May 15**, the 102nd NY Volunteers and 111th Pennsylvania Regiments, forming a forlorn hope, were ordered to charge up a hill occupied by the Confederates, and capture a battery interfering with the movements of Union forces. These were short but steep hills with narrow ravines. The 102nd NY, led by their gallant Colonel James C. Lane, marched up the hill in line of battle as steadily and in as straight a line as if on a dress parade, although all the way under a severe fire from the enemy. Immediately in front of the Confederate battery were rifle pits, occupied by a strong force, but after a severe fight, the rebels were driven out, and the battery, consisting of four guns, was captured and sent into Union lines. This brilliant operation was considered so hazardous that officers high in rank who saw it happening, expected both regiments would be annihilated. Colonel Lane of the 102nd wrote, "The 102nd moved in a line of battle down the hill and across a narrow valley and then charged in line of battle up the hill toward the enemy's fort. This movement was performed under a galling fire but the regiment was remarkably steady, not losing their alignment." WOTR Vol. 38 pt.2 p. 293

After its successful completion, Gen. Hooker, the Union army 20th Corps commander, personally complimented Col. Lane and his brave men for the splendid manner in which they had executed the work. Although it is debatable who won the battle of Resaca tactically, the Confederates withdrew when Sherman's men crossed the Oostanaula River, threatening Johnston's supply lines. Unable to halt the Union turning movement caused by Sherman's crossing of the Oostanaula, Johnston was forced to retire, burning the railroad span and a nearby wagon bridge in the early morning of **May 16**. After the Union repaired the bridges and transported more men over, they continued in the pursuit of the Confederates, leading to the Battle of Adairsville on **May 17**. The Battle of Resaca cost the Union roughly 600 dead and 3400 wounded and the Confederates roughly 3,000 casualties.

Confederate General Johnston's army retreated southward from Resaca while Union General Sherman and his armies pursued. Sgt. Moses Whitbeck and the 102nd NY Volunteers were on the march again. They immediately followed the Confederates in pursuit and bivouacked on the south side of the Coosawattee River a little after dark.

Failing to find a good defensive position south of Calhoun, Georgia, once across the Oostanaula River, Johnston sought to make a stand and draw the Federals into a costly assault. He expected to find favorable terrain near Calhoun, but in this he was disappointed and during the night of **May 16–17** led the Confederates southward toward Adairsville. Sherman followed, dividing his forces into three columns, and advancing on a broad front. There were skirmishes all along the route, but the main bodies were not engaged. Johnston continued to Adairsville while the Confederate cavalry fought a skillful rearguard action.

On **May 17, 1864** Union General Howard's IV Corps ran into entrenched infantry of Confederate General Hardee's corps, while advancing about two miles north of Adairsville. The 44th Illinois and 24th Wisconsin infantry regiments, led by Maj. Arthur MacArthur, Jr. (father of Douglas MacArthur), attacked Benjamin F. Cheatham's division and suffered heavy losses. Three Union divisions prepared for battle, but Union General Thomas halted them because of the approach of darkness. Sherman then concentrated his men in the Adairsville area to attack Johnston the next day. Johnston had originally expected to find a valley at Adairsville of suitable width to deploy his men and anchor his line with the flanks on hills, but the valley was too wide, so Johnston disengaged and withdrew toward Cassville, Georgia. Excerpt taken from New World Encyclopedia: Atlanta Campaign

When the Southerners abandoned Adairsville during the night of **May 17–18, 1864,** they withdrew across the Etowah River. As they fell back, their feelings were mixed. They had

lost a very strong position at Dalton, and had fallen back from
Resaca, Calhoun, and Adairsville. Now they were retreating
again under cover of darkness. The morning of **May 17**, as they
prepared for battle, their spirits had been high. Now their
disappointment was bitter. Although morale would revive in
the next few days, many Southern soldiers would never again
place as much confidence in Confederate General Johnston's
abilities as they once had. Johnston sent William J. Hardee's
Corps to Kingston, while he fell back toward Cassville with the
rest of his army. He hoped that Union General Sherman would
believe most of the Southerners to be in Kingston and
concentrate the bulk of his forces there. Hardee would then
hold off the Northerners at Kingston while Johnston, with
Generals Leonidas Polk and John Bell Hood, destroyed the
smaller Federal column at Cassville.

Sherman reacted as Johnston hoped, ordering James B.
McPherson and the bulk of George Henry Thomas's Army of
the Cumberland, which included the 102nd NY Volunteers
and Sgt. Moses Whitbeck, toward Kingston while sending only
John Schofield and one corps of Thomas' army along the road
to Cassville. Excerpts taken from MyCivilWar.com: The Battle of Adairsville.

On the morning of **May 19**, during the Union Army's
Atlanta Campaign, Confederate General Johnston ordered
General Hood to march along a country road a mile or so east
of the Adairsville-Cassville Road and form his corps for battle
facing west. While General Polk attacked the head of the
Federal column, Hood was to assail its left flank. As Hood
moved into position, he ran into Daniel Butterfield's Federal
brigade to the east. This was a source of great danger, for had
Hood formed facing west, these Federals would have been in
position to attack the exposed flank and rear of his corps. After

a brief skirmish with the Northerners, Hood fell back to rejoin Polk. Johnston, believing that the opportunity for a successful battle had passed, ordered Hood and Polk to move to a new line east and south of Cassville, where they were joined by Hardee who had been pushed out of Kingston. Johnston formed his army on a ridge and hoped that Union General Sherman would attack him there on **May 20**. As usual, the Southern commander was confident of repulsing the enemy.

Moses Whitbeck and the 102nd NY Volunteers, as part of the 20th Corps under Union General Hooker, were on the march once again, this time on their way to Kingston. Union generals Daniel Butterfield and Joseph Hooker, of the 20th Corps, were headquartered at the house of Confederate Colonel Hawkins F. Price, a state senator who had voted for Georgia secession in 1861. Hooker had been ordered from Adairsville to Kingston, on false reports that Johnston had retreated there. South of the Price house, Hooker discovered that Johnston had gone to Cassville. Excerpts taken from MyCivilWar.com: The Battle of Adairsville.

On **May 19, 1864**, the 4th Corps, followed by the 14th Corps, reached Kingston at 8 a.m. The 14th turned east to Cassville. A division of the 14th sent to Gillem's bridge over the Etowah River found no retreating Confederates. Johnston's forces were at Cassville, 5.5 miles east. McPherson's 15th Corps and 14th Corps, moving south from Barnsley's, camped on Woolley's plantation 2 miles west. The 20th Corps (which included the 102nd NY) were at Cassville.

That night, the Confederate leaders held a council of war. Exactly what happened at the council is a matter of dispute. According to Johnston, Polk and Hood reported that their lines could not be held and urged that the army retreat.

Believing that the fears of the corps commanders would be communicated to their men and thus weaken the army's confidence, Johnston yielded to these demands, even though he thought the position to be defensible. According to Hood, whose recollection of the council differs markedly from Johnston's, he and Polk told Johnston that the line could not be held against an attack but that it was a good position from which to move against the enemy. Johnston, however, was unwilling to risk an offensive battle and decided to fall back across the Etowah. No definite resolution of this dispute is possible, but most of the available evidence, including a post-war letter to Union General Butterfield, supports Hood's version of the conference. Certainly, Johnston was not required to allow the advice of subordinates to overrule his own judgment. The responsibility for abandoning the Cassville position rests solely on the Southern commander. Excerpts taken from MyCivilWar.com: The Battle of Adairsville.

Beginning **May 20, 1864,** the combatants of both armies took a brief respite to gather supplies and rest the troops. A soldier with a New York regiment recalled, "We were informed there would be no movement for several days. This would not only give us a good rest, but time to clean up and do our washing." The Atlanta Journal Constitution May 22, 2014

General Johnston and the Confederate Army withdrew from Cassville and on the evening and night of **May 20** and crossed the Etowah River and began heading in the direction of Allatoona Pass. General Johnston's line southeast of Cassville contained a flaw, a section of it, held by Lt. Gen. Leonidas Polk's Corps, was exposed to enemy artillery enfilade fire. That night, Polk and Lt. Gen. John B. Hood explained it to Johnston, saying that if the enemy attacked in the morning, they could not hold their position more than an hour. Johnston reluctantly agreed to order another nighttime retreat, this one across the Etowah River to high ground around Allatoona Pass, four miles south of the river. General Sherman decided to rest his army and they stayed on the north side of the river for the next three days, where they were resting and restocking supplies.

Having traveled through this area as a young officer, General Sherman knew the dangers of approaching the Confederate Army, at the heavily fortified Allatoona Pass. He formulated a plan to leave his railroad supply line and move southwest toward Dallas and try to out flank Johnston and take Marietta before the Confederate Army could get there.

Campaigning for less than three weeks, the maneuvers placed Union Maj. Gen. William T. Sherman approximately 45 miles closer to Atlanta; with each mile lost, anxieties increased for citizens across the "Empire State of the South." American Battlefield Trust: The Etowah River

Atlanta Mayor James M. Calhoun, upon learning of Johnston's movement across the Etowah, issued a call to arms, "In view of the dangers which threaten us … I require all the male citizens of Atlanta, capable of bearing arms (to report) to be organized into companies and armed."

The mayor ended his summons with a threat. Those men, he said, "not willing to defend their homes and families are requested to leave the city at their earliest convenience, as their presence only embarrasses the authorities and tends to the demoralization of others." Atlanta Journal Constitution: May 22, 2014

A period of relative quiet continued until **May 23**, when Sherman took a wide swing to his right, leaving his supply line behind in hopes of finally turning Johnston's flank. Sherman was content to let his forces rest for a few days.

May 20-23, 1864 Union General Sherman planned his next advance while he let his armies rest and resupply. He knew the defensive strength of Allatoona, Georgia, and was not going to test it with an infantry assault. Instead he planned to cross the Etowah River downstream, and march his columns south toward Dallas, a crossroads village southwest of Johnston's position, flanking it and threatening his rear.

After the Cassville fiasco, the Confederates continued their retreat to the Etowah River, crossing the river, burning the railroad bridge behind them, and taking up defensive positions south of the river in the rugged Allatoona hills. Aware of the character of these hills from a personal visit many years before, Sherman chose to suspend his direct advance along the railroad. Placing his army along the north side of the river westward some ten miles to Kingston, Sherman set about devising a new ten-day strategy that would carry his army directly southward through the tangled wilderness countryside. He would temporarily abandon his rail line in favor of a mule-drawn wagon alternative. Surely a risky strategy, he reasoned, but it would force Confederates to abandon their impregnable Allatoona forts to protect their flanks and railroad. With a head start, he might even beat his opponent to those country crossroads near Dallas and New Hope Church – roads that led directly to Atlanta, bypassing the formidable fortress now being constructed at Kennesaw Mountain in Cobb County. The complete dependence on wagons and mules to supply his immense army in this landscape of bad roads and worse maps was a bit unnerving. Still, the risk seemed reasonable and the goals worth it. The move could certainly offer an opportunity to choose his fields of battle, and perhaps even gain a rapid and politically timely capture of Atlanta. American Battlefield Trust: Cobb County in the Atlanta Campaign

May 24, 1864 The 102nd NY Volunteers and Sgt. Moses Whitbeck were on the March again. They left Cassville, Georgia, on the day before and crossed over the Etowah River on a pontoon bridge just south of the mouth of Euharlee Creek and the whole 20th Corps camped along the creek that night. On this day, they moved forward to Burnt Hickory or Huntsville and camped there. There was some skirmishing going on between them and the rebels, but it is unclear if the 102nd was involved. That evening, they had a tremendous storm of thunder and lightning that lasted most of the night and as most of their wagons had not caught up with the corps the men spent the night where they could in their "rubbers."

By **May 25, 1864**, having waded the Etowah River several days earlier, the various Union marching columns were converging rapidly on Dallas, Georgia. After Johnston retreated to Allatoona Pass on May 19–20, Sherman decided that he would most likely pay dearly for attacking Johnston there, so he determined to move around Johnston's left flank and steal a march toward Dallas. Johnston anticipated Sherman's move and shifted his army into Sherman's path, centering a new line at New Hope Church. Sherman mistakenly surmised that Johnston had a token force and ordered Maj. Gen. Joseph Hooker's 20th Corps to attack.

Union General Joe Hooker guided his 20th Corps along a wagon path leading toward a place known as Owens Mill, on a creek called "Pumpkinvine." He hoped to reduce his march time to Dallas by avoiding the congestion of military traffic on the main routes.

Union General John Geary, commanding Hooker's Second Division, which included Sgt. Moses Whitbeck and the 102nd NY Volunteers, led the way. Approaching the mill, Geary received rifle fire from the ridge beyond; a brigade was deployed across the creek to drive off the Confederate skirmishers. These first few shots signaled the beginning of the Battle of New Hope Church. The Union soldiers in this battle would come to call the crossroads at New Hope Church the "Hell Hole." Advancing with his three divisions in parallel routes, Hooker pushed the Confederate skirmishers back for three miles, before coming to Johnston's main line.

Geary, alarmed by the aggressive manner of the small band of Confederate skirmishers in his front, concluded, therefore, that a large Confederate force must lie just beyond. He postponed his attack three hours while awaiting the arrival of Butterfield's and Williams's divisions. It was almost 5 p.m. before the three Union commands began their movement toward the New Hope Church crossroads.

Things went downhill from the beginning. First, a terrific thunderstorm with frequent lightning and heavy downpours of cold rain set in shortly after 5 p.m. Complicating this, the Union commanders decided upon an ill-chosen formation, a column of divisions by brigade. Such a formation, while improving command control, exposes the flanks of the approaching column to rifle crossfire while offering very little opportunity to return such fire. At New Hope Church, this formation had the effect of negating a three-to-one Union numerical advantage. An understrength division of Confederate infantry in position at the church cemetery, without earthworks, had little difficulty stopping the attack 300 yards short of its objective at New Hope Church. The assault ground to a halt by nightfall. The night was a disaster of confusing orders, pitch black darkness, cold rain, and desperate men entrenching in ravine-laced wooded thickets strewn with the wounded and dead. That night, the place truly earned its long remembered epithet, "The Hell Hole at New Hope Church." Having suffered nearly 2,000 casualties at New Hope Church, and with the fast-marching Confederate infantry already covering the key Atlanta-bound roads near Dallas, Sherman reviewed his options. American Battlefield Trust: Cobb County in the Atlanta Campaign

On **May 26, 1864**, frustrated by the tactical disappointments but also by the increasing supply shortages, Union General Sherman decided to abandon his "wilderness" plan in favor of returning to the supply security of the more dependable railroad. First, though, he would try to locate the east flank of the Confederate earthworks.

The Fourth Corps of the Cumberland army was chosen for the task. Should an attack be possible, the Fourth Corps was to be supported by units from the Army of the Ohio and the Fourteenth Army Corps. The battle at Pickett's Mill began shortly after 4:30 p.m. As at New Hope Church two days before, it was again a tactical disaster for Sherman. In that blind wilderness, three good brigades from the Fourth Corps suffered nearly 2,000 casualties and gained no great advantage. A witnessing Union officer described this mismanaged attack as "The Crime at Pickett's Mill." Sherman curiously made no mention of the battle in his official report, nor still later, in his memoirs.

But, Union Brigadier General Alpheus Williams wrote much about the battle at New Hope Church in his letters to his daughters. He described that he lost about 800 men killed

and wounded from his 1st Division of the 20th Corps. He also described that after five days his division had not advanced an inch and that many of their dead lay unburied. The rebels were well entrenched in dense woods with considerable underbrush.
Excerpts taken from American Battlefield Trust: Cobb County in the Atlanta Campaign

General Williams wrote, "The Rebs. Have evidently a strong place and I suppose have collected all their forces in the South to give us a final meeting... I suppose we shall move somewhere soon. It is a very tedious and worrying life as we are situated, for we are kept constantly on the qui vive ready for battle. Our rest, you can well fancy, is not of the most refreshing kind." FTCM p. 314

General Geary, who commanded the Second Division that included the Third Brigade and the 102nd NY Volunteers and Sgt. Moses Whitbeck, wrote, "It is now 26 days since we left Bridgeport, nine of which has been in actual battle. How our present engagement will terminate is in God's hand, undecided to mortal eye. We are now 5 days under fire consecutively." APGTW p. 177

May 27, 1864. The Union Army at New Hope Church and Picket's Mill remained entrenched and waited for orders from General Sherman. The dead and dying laid all around in the thick wilderness and underbrush that did not allow either side to move.

Stopped in his tracks at New Hope Church on May 25-26, Maj. Gen. William T. Sherman was determined to circumvent Gen. Joseph Johnston's army once again by sending Maj. Gen. Oliver O. Howard's IV Corps on a flanking move, this time around the Confederate right. After a five-hour march through rugged terrain, Howard's men found the rebels entrenched near Pickett's Mill, 15 miles west of Marietta. On the afternoon of May 27, Howard ordered an assault against the Confederate works. The Confederates were ready for the attack, which did not unfold as planned because supporting troops never appeared. Waiting were 10,000 Confederate troops under Maj. Gen. Patrick Cleburne. The Federal assault began at 5 p.m. and continued into the night. Dismounted Confederate cavalry harassed the left flanks of the six attacking Federal brigades. The Confederates repulsed the attack causing high casualties. Daybreak found the Confederates still in possession of the field. The Confederate victory emboldened Johnston.

Author Ambrose Bierce fought for the Union at Pickett's Mill as a topographical engineer under William Babcock Hazen. Bierce's short story "The Crime at Pickett's Mill" is about this battle. Bierce reported that the battle took about 45 minutes, that the total loss was 1,400 men with one-half were killed and wounded in Hazen's brigade in 30 minutes of actual fighting. The 20th Corps with the 102nd NY Volunteers and Sgt. Moses Whitbeck did not engage this day but were entrenched near Picket's Mill. Excerpts from American Battlefield Trust: Pickett's Mill Battle Facts and Summary

May 28, 1864 marked the Confederate's turn for tactical mistakes. Suspecting correctly that Union General McPherson's Army of the Tennessee at Dallas, Georgia, was preparing to shift eastward toward New Hope Church and the railroad, Confederate General Johnston instructed Lt. General Hardee to attack immediately should he detect any such movement. Hoping to catch the Union infantry off balance while in the act of shifting position, Hardee ordered Bate's division to carry out the task by striking simultaneously at two geographically separated points, one at the Villa Rica Road south of Dallas, and the other a mile distant on the Marietta Road east of the village. Curiously, the signal to attack was to be the sudden sound of heavy gunfire at the south end of the line. Confused by gunfire from a related cavalry action near the Villa Rica Road, a Kentucky unit on the Marietta Road prematurely launched a full-scale assault. This unit, the famed Kentucky Orphan Brigade, was so badly mauled by being caught in a crossfire that it was subsequently disbanded. Bates' division became entangled with the enemy and had difficulty disengaging, requiring Hardee to commit additional units from his command to the rescue. The battle continued for the better part of the day along the entire Dallas line. Confederate casualties on this day are estimated to have exceeded 1,000 men.

Sherman's continued movement eastward to the railroad eventually placed McPherson near the south bank of the Etowah River where he set about repairing the river bridge and the rail tracks south toward Acworth. Meanwhile, the Cumberland and Ohio armies entered Cobb County from the west following the Stilesboro and Burnt Hickory Roads and colliding almost immediately with a formidable 12-mile line of fortifications known as the "Lost Mountain/Brushy Mountain Line." Excerpts taken from American Battlefield Trust: Cobb County in the Atlanta Campaign

May 29, 1864, even though the rebels were repulsed the previous day by Maj. Gen. John A. Logan's 15th Corps and Maj. Gen. Grenville Dodge's 16th Corps, they stayed in the area and there was much skirmishing in the Dallas and Pickets Mill area. Light skirmishing devolved into heavy fighting, with each side pouring more men into the struggle . The rebels were eventually repulsed, though both sides suffered heavy casualties. Both armies remained around Dallas and skirmished heavily for three days. General Geary wrote in his report, "During the days of the 28th, 29th, 30th and 31st of May, our position and daily routine of artillery practice and sharp-shooting were unchanged. Small out-works for the protection of skirmishers were thrown up at every available point, thus diminishing my daily return of casualties. From the 25th of May until the 1st of June my entire division was under fire, without an hour of relief. Owing to the proximity of the lines, and the nature of the ground, no one, whether in front or rear, could rest quietly with any assurance of safety. No opportunity being afforded for proper shelter, rest, and diet, the necessary result of this series of operations was a large increase of sickness." WOTR Vol 38 pt. 2 p. 125

Union General Sherman, looking to outmaneuver Johnston again and resupply his army, left the area on June 1 and moved his army back to the Western and Atlantic Railroad. For the rest of June, the armies fought a series of skirmishes around Marietta, just 20 miles north of Atlanta. Excerpts taken from American Battlefield Trust: Dallas Battle Facts and Summary

But something happened that severally threatened the Whitbeck family. Moses Whitbeck was severely wounded near Dallas, Georgia, on **May 29**. He suffered a severe gunshot wound to the right side of his head while bearing a rifle. General Geary commanding the Second Division of the 20th Corps, and thus the commander of the Third Brigade that held the 102nd NY Volunteers would write that his command lost about 1200 men killed, wounded or missing. Colonel Van Buren of the 102nd NY Volunteers reported, "At 6 p.m. of the 26th we moved the distance of the front of our brigade toward our left, where we remained until 7 a.m. 31st. Here we lost in all 2 enlisted men killed and 19 wounded." WOTH Vol 38 Part 2, p. 294 Moses Whitbeck was one of those wounded.

Moses was moved by wagon to Kingston, Georgia, and then

he was sent by train to Chattanooga, Tennessee, then hospital-
ized at Chattanooga, before being transferred to Alexandria,
Virginia, for his recovery and eventual return to the 102nd NY
Volunteers. Praise God for saving Moses because if he hadn't,
I would not be writing this history post today. Thank God for
Surgeon Kendall of the 149th NY Volunteers and Surgeon
Applegate of the 102nd NY Volunteers for caring for Moses
Whitbeck in the field and getting him ready for transport to
Chattanooga.

June 1, 1864 The 20th Corps of the Union Army of the
Cumberland left their position near Dallas and marched about
four miles northeast and took position on a hill called
Brownlow Hill. From there the country was woods and
mountain ranges as far as the eye could see. To the East lay the
Kennesaw Hills near Marietta and to the South East was the
solitary Lost Mountain. Here they would stay for the next four
days and there were many meetings between the Union
Generals to plot their next move. As for Moses Whitbeck, he
was on his way to a hospital in Chattanooga, where he would
recover from the serious head wound.

Elements of Sherman's Army began arriving in Cobb
County, Georgia, on **June 2, 1864**.

June 3, 1864 Daily rains began and lasted for two weeks.
Wagons and artillery caissons were soon buried to hubcaps, as
all traces of roadways disappeared in seas of mud. Wheeled
vehicles would use the "road." Infantrymen, required to slog
through the fields alongside the roads, endured the misery of
rough terrain and briar patches, with full exposure to swarms
of ticks, red bugs, and mosquitoes. Tents pitched between
puddles of water at an evening campsite, amid downpours of
rain and swarming insects, promised little rest after a miserable

day. The 102nd NY Volunteers, now without Sgt. Moses Whitbeck, were ordered to move and take possession of a bridge
across Allatoona Creek and within three miles of Acworth,
which they accomplished at 9 p.m. even though they were in a
drenching rainstorm. Along with the 78th NY Volunteers, the
102nd acted as flankers for the brigade. General Geary, who
commanded the division that the 102nd was part of, wrote,
"This is the evening of the 11th day's hard fighting and we are
now in possession of Allatoona and Acworth on the east side
of the Allatoona Mountains. The total casualties of my
Division number over 1200 men (Moses Whitbeck was one of
them). I am happy to state that my glorious old Division has
maintained its previous prestige… We are getting along very
slowly but surely. The country is almost unbroken forest."
APGTW p. 177-178

As the Cumberland and Ohio Armies entered Cobb
County, they collided almost immediately with the Lost
Mountain/Brushy Mountain Line.

In early **June 1864**, the battle line was occupied near Lost
Mountain by the four divisions of Confederate General Hardee's corps, with the three divisions of Polk's Corps being next
in line eastward. Hood's three divisions manned the Brushy
Mountain third of these fortifications. Union General Sherman
knew the Confederate infantry was too small to sufficiently
cover such a distance. Hardee's men occupied the works westward only as far as the Burnt Hickory/Sandtown road intersection (Acworth Due West Road today) His left was anchored near the log church of Gilgal. The defense of the mile or so
of trenches westward to Lost Mountain became the responsibility of Confederate cavalry. Excerpts taken from American Battlefield Trust:
Cobb County in the Atlanta Campaign

By **June 5, 1864**, Hardee had moved Bates's division a mile north to the crest of Pine Mountain, a hill overlooking Thomas's advance along the Stilesboro Road. Polk's Corps slipped a division length westward to link with Hardee, covering the absence of Bates in the battle line.

On **June 6, 1864**, the 102nd NY Volunteers were on the march again, heading to within eight miles of Marietta. Once there, they constructed breastworks where they would remain until June 12th.

June 7, 1864 The Union Army of the Cumberland 20th Corps, which included the 102 NY Volunteers in the Second Division under Brigadier General John Geary, had been on the march the previous day toward Marietta, Georgia and camped at Hull's farm near the junction of the roads leading to Big Shanty and Lost Mountain. Here they would remain until June 12th, building breastworks so they could cover these roads and send skirmishers looking for Confederate lines. They discovered that the rebels were fortifying the ridges connecting Lost Mountain and Pine Hill.

On **June 8, 1864** Union General Geary wrote a letter to his wife in which he stated, "My troops are in line of battle, laying upon their arms. The enemy are also in line, a little beyond the reach of our cannon, and tomorrow may possibly bring about a collision of arms. We had truly a victorious march and have pushed the enemy more than one hundred miles, still he is not whipped sufficiently to relieve him of his arrogance."

"…As we pass through the country, we leave it as though all the locusts of Egypt had been upon it. There is not a single blade of grass left upon the earth. Wheat fields are eaten to the ground, and the rising corn is beginning to yield its quota to the sustenance of our animals. The provisions of the people is also taken without compunction and they are left in utter want." APGTW p. 179

June 12, 1864 The rain that part of the year in the Atlanta and Marietta area had been mostly constant. This had made the roads where the Union and Confederate Armies were positioned very muddy and mostly unusable. So the two armies hunkered down for the most part, skirmishing and probing enemy lines to determine strength and position. General Williams of the 20th Corps First Division wrote, "The weather which had been for some days of alternating heavy showers and sunshine became a heavy northeast storm and lasted without intermission for two or three days, so cold that with big camp fires and overcoats one was hardly comfortable. The earth became saturated like a soaked sponge and the mud was intolerable." FTCM p. 320

The 102nd NY Volunteers was ordered three miles to the rear of the Union Army lines where they guarded their trains and wagons. Here they received some relief from the constant skirmishing. Moses Whitbeck was in a Chattanooga hospital recovering from his head wound.

On **June 13, 1864**, after having been moved away from the front the previous day to guard trains and wagons in the rear of the Union Army lines, the 102nd NY Volunteers were ordered to the front of the Union lines where they rejoined the Third Brigade that they were members of, and built breastworks in the rear of the main line opposite Bald Mountain. The artillery batteries of the 20th Corps were placed on a ridge 800 yards from a prominent knoll called Pine Hill.

On **June 14, 1864** near Kennesaw Mountain, the 102nd NY Volunteers took up a position to the right of the 29th

Pennsylvania Volunteers and built breastworks. George Stilwell of the 102nd New York wrote his brother telling him of the heavy cannonading he heard that day. "Fights are getting to be a common thing now days," he wrote, likely unaware that a Federal artillery shell had not only killed one of the Confederacy's most popular generals but will also lead to a heavy fight for him and his comrades in Brig. Gen. John W. Geary's "White Star" division of the 20th Corps. That day, Confederate Generals Hardee, Johnston, and Polk met at Pine Mountain, concerned that Confederate General Bates's division was becoming isolated by Union movement near its flanks. Atop the mountain, the meeting ended tragically with popular General Polk's death by an artillery projectile. A Union gun a mile away near the Stilesboro Road had fired the chance shot. (Today a well preserved four-gun earthwork marks the site of that Union battery. Southward at the crest of Pine Mountain, infantry and artillery entrenchments share the location with a granite shaft memorializing the place of Polk's death.)

Since the Stilesboro and Burnt Hickory Roads run roughly parallel in a southeasterly direction toward Kennesaw Mountain and Marietta, it is not surprising that Union armies using these roads frequently engaged in joint military actions during the first two weeks in June. One such occasion occurred on June 15 at the Battle of Gilgal Church/Pine Knob. The purpose of this attack was to probe and possibly break the overextended Confederate battle line, forcing a precipitous retreat. Excerpts taken from American Battlefield Trust: Cobb County in the Atlanta Campaign

By mid afternoon on **June 15, 1864,** Daniel Butterfield's division approached the Burnt Hickory/Sandtown roads intersection intending to strike the Confederates entrenched at the crossing. Sherman had chosen Hooker's 20th Army Corps for the task with units from the Ohio army protecting the 20th Corps' right flank. Planned as a coordinated assault by Hooker's three 5,000-man divisions (Butterfield's, Geary's, and Williams's) on a mile-wide front extending from Gilgal eastward to Pine Knob, much depended on timing and interdivisional communication, neither of which would be forthcoming.

Judging that Butterfield had gotten in position a mile west on the Sandtown Road, Geary and Williams at the foot of Pine Mountain began their advance southward around 5 p.m., guiding on a distant wooded hill they called Pine Knob. With Geary's men leading, including the 102nd NY Volunteers, and Williams's following closely, they began a ridge-by-ridge struggle toward Pine Knob, the hill believed to mark the location of the Confederate battle line.

Upon contact with the principal Confederate defenses, Williams's division was to slide westward, protecting Geary's right and tying with Butterfield's left, thus forming a united three-division battle front for the final assault. Ridge-creased terrain, stubborn resistance by Confederate skirmishers, and approaching nightfall defeated the plan. The final assault never materialized. Today, a 20-acre wooded tract with earthworks marks Butterfield's battleground at Gilgal; one mile east, a 5-acre history preserve with a historical marker locates the forward-most position gained by Geary that night at Pine Knob. Sherman's casualties in this failed effort are estimated to have been just fewer than 1,000 men. That same day a diversionary attack by Union General McPherson's Corps at the foot of Brushy Mountain was more successful, netting the capture of some 300 Alabama infantry. Tactically, it was the one bright spot in Sherman's otherwise rather dismal day. One bad spot for the 102nd NY was that Colonel Lane became sick and Major Lewis Stegman took command and at 3pm he led them in their charge up Lost Mountain, skirmishing in a heavy line with rebel forces, where their advance was halted and the men had to stay in place overnight. The men were worn out and exhausted by continual marching, building breastworks and under continual fire since May 25, the roads had been heavy from 10 days of rain and had just fought a six-hour battle. Excerpts taken from American Battlefield Trust: Cobb County in the Atlanta Campaign

On **June 16, 1864**, the 102nd NY, unable to progress the day before because of heavy skirmishing, fell back a short distance to breastworks that were made the night before by the rest of the Third Brigade. Major Stegman and Lieutenant John R. Elliot were both wounded in the leg. The 102nd NY lost 4 enlisted men killed and 14 wounded, and Captain Barent Van Buren was put in command.

The tactical defeat on the **June 15, 1864**, evolved into a strategic advantage for Sherman two days later. Learning that Confederate cavalry had abandoned their trenches toward Lost Mountain, leaving his flank exposed to enfilading artillery fire from the Army of Ohio, Hardee withdrew several miles to the east bank of Mud Creek the night of June 16. Here he anchored his right on a steep hill (now called French's) tying to the left of French's division of Polk's Corps (now commanded by Loring), forming here a pivot or salient. Hardee's left would simultaneously swing south two miles to a position along the east bank of Mud Creek to a point just beyond the Dallas/Marietta road. Thus Hardee substantially reduced the length of his front and better protected his flanks. This new alignment of fortifications became known as the "Mud Creek Line." American Battlefield Trust: Cobb County in the Atlanta Campaign

On **June 17, 1864** at the Mud Creek Line, in a sudden dash during a thunderstorm, three regiments led by one-armed Colonel Frederick Bartelson (wounded at Chickamauga and a recently returned prisoner of war from Libby Prison) captured a position near French's Hill. Equipped with Spencer repeating rifles, they succeeded in holding the point through the night despite several Confederate counterattacks. Bartelson's location posed a serious threat to the new Confederate line of defense. The 102nd NY followed the rebels about 2 miles where the men once again built breastworks and slept on their arms.

On **June 18, 1864**, Confederate General French's division was pounded by a day-long crossfire of Union artillery. On the same day, at the Dallas/Marietta road near the Darby House, Confederate General Hardee's anchor fort was destroyed in an intense three-hour duel with two Union batteries attached to the advancing Army of the Ohio. The recent rainstorms continued, and the armies keep skirmishing, as it was hard to move about because of all the mud. Also the artillery batteries of the Union were hammering the Confederates and General Williams who commanded the First Division of the 20th Corps wrote an interesting story in one of his letters to his daughters about how the rebels were scrambling around and were disorganized because of the shelling they were receiving from Union Batteries. He wrote, "Suddenly we saw, a half a mile or more on our right, a great cloud of rebel cavalry flying in disorder to the rear. There must have been a brigade of them and every man was kicking and spurring for dear life. Many horses were riderless. We opened on them with artillery, which greatly increased the disorder. It was laughable to see (General) Hooker's excitement. 'Williams,' he would cry out, 'see them run. They are thicker than flies in a Mexican ranch' 'See them go,' and we all shouted… The rain came down in torrents all the afternoon and night and the mud seemed too deep to ever dry up." FTCM p. 321

General Geary, who commanded the Second Division of the 20th Corps that the 102nd NY Volunteers were part of wrote, "It still rains incessantly and the roads are wonderful (sarcasm). The enemy are becoming more and more desperate." APGTW p. 182

In the predawn hours of **June 19, 1864**, in the Union Army's drive toward Atlanta the Mud Creek battle line would be abandoned. The Confederates withdrew to the foothills of Kennesaw Mountain. Following much rain and skirmishing by both Armies, the Union had been shelling the confederate positions and driving them back and closer to Atlanta. Both sides had been receiving many casualties and there was much action, but historians largely gloss over what happened and just say that the Union drove on Atlanta to capture it. In reality, there was a very large amount of fighting happening in all these "small" battles. They just have not been given the notoriety that the "big" battles received. The 102 NY Volunteers as Part of the 20th Corps and the brigades and divisions of the three Union Armies were a large part of all these maneuvers. Since Moses Whitbeck was injured in the head on May 29 and sent to a hospital in Chattanooga, Tennessee, he was obviously not involved in these latest rounds of fighting.

On **June 20, 1864**, near Marietta, the Josiah Wallis House on Burnt Hickory Road became the headquarters of Union

General Oliver O. Howard, commander of the Fourth Corps. Here, for the next few days, Howard directed the attacks on nearby hills – including division-size assaults on hills lying behind the knee-deep swampy flats of Noyes Creek. One of these hills would later be called "Nodine's" following heavy combat there. Kirby's and Nodine's brigades gained and lost the hill several times on June 20. The 102nd NY as part of the 20th corps, were relieved that day and sent to the rear of the Union lines for some rest. Under direct orders from an impatient Howard, the two brigades of the IV Corps tried again, succeeding this time in holding the ground despite rebel counterattacks and heavy concentrated barrages of artillery. The fighting here on this and nearby hills in mid-June was especially close and personal, combative and aggressive – often hand-to-hand and frequently continuing into the night. Excerpts taken from American Battlefield Trust: Cobb County in the Atlanta Campaign

June 21, 1864 The 102nd NY were ordered further to the right of the Union lines near Kennesaw Mountain and built breastworks around the area of Noses creek. General Geary's troops of the Second Division of the Union Army 20th Corps which included the 102nd NY Volunteers were probing the Confederate left to find a way to flank Confederate General Johnston and take Murrieta. They were facing rebel forces lead by Generals Hardee and Hood, and they were skirmishing as they went.

June 22, 1864 Confederate General Johnston had withdrawn the Army of Tennessee to a defensive position astride Kennesaw Mountain near Marietta. Initially, Union General Sherman decided that Johnston's new line was too strong to risk a frontal attack. Instead, he ordered Major General John Schofield, commanding the Army of the Ohio, and Major

General Joseph Hooker, commanding one corps of the Army of the Cumberland, to extend the Union line west and attempt to turn Johnston's left flank.

Anticipating Sherman's plan, Johnston countered by sending Lieutenant General John B. Hood and one corps of his army to reinforce his left flank. By June 22, Hood's soldiers were in position, and one of his divisions encountered two federal regiments near Kolb's Farm. Underestimating the number of Yankees in the area, Hood launched an attack. Warned of Hood's presence in the area, Union Generals Hooker and Schofield prepared for the Confederate assault by entrenching their soldiers. This included the 102nd NY Volunteers. Hampered by swampy terrain, Union artillery, and Hood's poor reconnaissance, the Yankees repulsed the rebels inflicting high casualties. The 20th Corps would stay in place until June 27.

Confederate casualties at the Battle of Kolb's farm more than tripled federal losses. The Confederacy suffered about 1,000 casualties (killed, wounded, captured/missing) compared to roughly 300 losses for the Union (killed, wounded, captured/missing). Although the Union forces won the Battle of Kolb's Farm, they failed to achieve their original objective of turning the Confederate flank. Excerpts taken from AmericanHistory Central.com: Battle of Kolb's Farm, 1864

Moses Whitbeck returned to corporal on **July 10, 1864.** There is no information about this other than it happened and is in the War Records.

July 12, 1864 By order of the War Department, the 78th NY was consolidated with the 102nd New York, and from that time its honored name ceased to appear in the records of the campaign. Some of the officers were retained in service, with

Hammerstein and Chatfield taking command of the consolidated regiment without loss of rank. The subsequent history of the 78th NY is the same as that of the 102nd NY, which served with honor and distinction until the end of the war.

Colonel Hammerstein of the 78th reported, "Though its name is lost, its services will be remembered by history, and its brave survivors are still in the field ready and willing to do battle for the Union to the end of the rebellion." WOTR Vol 38 P. 290 Colonel Hammerstein became the Commander of the 102nd NY. Colonel Lane was discharged this day according to war records because of the wounds he received at the Battle of Gettysburg.

July 17, 1864 The 102nd NY Volunteers, having been bolstered by the addition of the 78th NY survivors broke their camp at the north side of the Chattahoochee and marched to the left as far as Pace's Ferry, crossed the Chattahoochee and marching a short distance further bivouacked for the night.

July 18, 1864 The 102nd NY left Pace's Ferry about 2 p.m. and crossed Nancy's creek, marched about four miles on the Decatur Road and halted not far from Howell's mill.

July 19, 1864 The 102nd NY marched early in the morning on the road toward Atlanta until the regiment had reached Peach Tree Creek. Toward evening the regiment moved to the Union left and was placed near and in support of the batteries placed near the road crossing the creek. A portion of the 149th NY was ordered to cross the creek and forming a line of skirmishers and advance into and through the woods to the opposite side of the creek, occupied by the enemy. To cover their advance, Colonel Hammerstein was ordered to keep up a heavy firing by file. This was done until rendered unnecessary by the rapid advance of the 149th NY Volunteers, led by

Colonel Barnum and Lieutenant-Colonel Randall. The 102nd NY Regiment soon after crossed the creek and taking position on the left of the 111th Pennsylvania Volunteers threw up a line of earth works and rested for the night.

July 20, 1864 The Battle of Peachtree Creek was the first major assault by Lt. Gen. John Bell Hood since taking command of the Confederate Army of Tennessee. The attack was against Maj. Gen. William T. Sherman's Union Army, which was perched on the doorstep of Atlanta. The main armies in the conflict were the Union Army of the Cumberland, commanded by Maj. Gen. George Henry Thomas, and two corps of the Army of Tennessee, which included the Second Division of the 20th Corps which the 102nd NY Volunteers were assigned to, and were right in the middle of the Union lines and were pressured hard by rebel forces. Peach Tree Creek was about four miles from Atlanta, and the Confederates were desperate to hold it.

Hood's intention was to attack Union forces there before they could cross the creek. The Union forces began preparing defensive positions, but had only partially completed them by the time the Confederate attack began. After three and a half hours of hard fighting, the Second Division was able to defeat the Confederates. The Union lines had a tough time coordinating their attack because of the difficulties of the ground and terrain and were actually driven back at one point by the rebels. However, the Thirteenth New York artillery battery, which was considerably in advance on the left of the brigade, and the 60th NY Volunteers and the 29th Pennsylvania Volunteers had been placed in position on the left and right angles to the line of battle of the brigade in support of the battery. Their steady and destructive fire upon the flank of the advancing enemy, stopped his advance, and taking advantage of this, the brigade was rallied and made good their stand against greatly superior numbers, and finally forced back the enemy's columns in disorder.

The battle ceased at dark, works were hastily erected, and the men laid on their arms that night. The Union lines had bent but were not broken under the weight of the Confederate attack, and by the end of the day the rebels had failed to break through anywhere along the line. Hood withdrew into defenses of Atlanta the following day. Estimated casualties were 4,250 in total, 1,750 on the Union side and at least 2,500 on the Confederate side. Geary's division suffered greatly and had 476 casualties, 28 percent of all the Union loses.

July 21, 1864 The Union Army paused for a day in their drive toward Atlanta to bury the dead from the Battle of Peach Tree Creek the day before and to attend to those who were wounded. For the 20th Corps, the Atlanta campaign had been hard and wearisome. For more than a month, they had been under fire literally day and night. Since the Union crossed the Etowah River, the Confederates had entrenched themselves every five miles. Driven from one line, they would fall back to the next line and each one seemed stronger than the previous one. The country was all woods, deep ravines, muddy creeks and steep hills, very defensible positions by nature. The 20th Corps, which included the Second Division and the 102nd NY Volunteers, were engaged throughout the campaign and earned the nickname the "iron clads." Many requests were made from dozens of regiments and some brigade commanders asking transfer to the 20th Corps.

July 22, 1864 The Battle of Bald Hill east of Atlanta (sometimes called the Battle of Atlanta) on the Georgia railroad occurred. Confederate General Hood attacked the left wing of Union General Sherman's forces, the Army of the Tennessee under James B. McPherson. During the fighting, McPherson had gone to check a gap in his line when Confederate forces emerged from the brush and demanded his surrender. McPherson doffed his hat like he was going to surrender, but then turned and tried to flee, but he was shot by rebel forces. The 102nd NY as part of the Third Brigade in the Second Division of the Union Army 20th Corps advanced about four miles and came up toward the rebels and their well-constructed earthworks to the west of Atlanta. The Brigade quickly set up their own earth works and bivouacked there for the night. This bloody day of fighting accomplished little, but cost Hood about 5500 out of 35,000 troops. The Union lost about 3,700. Some historians claim that the 20th Corps, of which General Geary commanded the Second Division that included the 102nd NY, missed an opportunity to storm the city of Atlanta while this fighting went on.

Beginning **July 23, 1864**, Union General Sherman settled into a siege of Atlanta, shelling the city and sending raids west and south of the city to cut off the supply lines from Macon, Georgia. Both of Sherman's cavalry raids, including McCook's raid and Stoneman's Raid, were defeated by Confederate cavalry collectively under General Wheeler. Although the raids partially achieved their objective of cutting railroad tracks and destroying supply wagons, they were soon after repaired, and supplies continued to move to the city of Atlanta. Following the failure to break the Confederates' hold on the city, Sherman began to employ a new strategy. He swung his entire army in a broad flanking maneuver to the west. Finally, on August 31, at Jonesborough, Georgia, Sherman's army captured the railroad track from Macon, pushing the Confederates to Lovejoy's Station. With his supply lines fully severed, Hood pulled his troops out of Atlanta the next day, September 1, destroying supply depots, as he left to prevent them from falling into Union hands. He also set fire to eighty-one loaded ammunition cars, which led to a conflagration watched by hundreds. As for the 102nd NY Volunteers they continued to add to their breastworks in the morning and in the afternoon they marched to the left a short distance and relieved two other regiments of the Third Brigade, occupying their works. Here they remained under fire, but received no real damage. Here they stayed until July 26 and strengthened their position. Excerpts taken from Wikipedia: Battle of Atlanta.

July 27, 1864. Maj. Gen. Oliver O. Howard, U.S. Army, assumes command of the Army of the Tennessee. The 102nd NY was part of the Army of the Tennessee in the 20th Corps, Second Division. Union General Alpheus Williams from the 20th Corps First Division takes charge of the 20th Army Corps in place of General Hooker on the 28th. Hooker, partially because Union General Howard was junior to him and partially because he blamed Howard (who had commanded XI Corps at the Battle of Chancellorsville, where it had been routed during Stonewall Jackson's famed flank march) for his part in the defeat of the Army of the Potomac, resigned. Seeing the Civil War.com: 20th Corps

Union General Sherman chose Major General Henry W. Slocum to replace Hooker, but he was in Vicksburg and General Williams will be the Commander until General Slocum arrives. Also on this date the Artillery of the 20th Corps was detached from the corps and was organized as a brigade under Major Reynolds, First New York Artillery. The whole Corps was in trenches in Front of Atlanta, Georgia, occupying 2.75 miles of the line until August 25.

July 28, 1864 The position of the Union Army 20th Corps was near the Western and Atlantic Railroad, about two miles from the Center of Atlanta and extending north. This Union line would remain to besiege the city and at times parts of the line moved slightly forward and other units then moved to perfect the line. The firing of the light guns of the Corps were kept up daily and several times the Union feigned attacks but did not advance. Here the 102nd NY Volunteers would remain until August 26, strengthening their works and keeping alert for any movement which the enemy might attempt.

August 25, 1864 The Union 20th Corps was pulled back to the Chattahoochee River to hold the crossing places and guard the railroad communications while the balance of the army operated south of Atlanta. The Second Division and thus the 102nd NY Volunteers were near the Howell's Mill and Pace's Ferry road. They were directed to occupy the high ground above the military bridge at Pace's Ferry with two brigades and to send one brigade to the prominent bluff about a mile north of the railroad bridge at Montgomery's Ferry.

August 26, 1864 The Second Division of the 20th Corps under General John Geary reached Pace's Ferry about 4 a.m. having been held up by the 4th Corps the night before. The Third Brigade which included the 102 NY Volunteers was posted on the left extending across the Buck Head road covering the bridge at the ferry. That afternoon rebel cavalry skirmished with them but were driven back but remained busily engaged probing the Union lines.

August 27, 1864 Maj. Gen. Henry W. Slocum, U.S. Army, assumes command of the 20th Army Corps. The corps were perfecting their lines and building breastworks and skirmished with rebel troops who came forward to probe their position. From the 27th until September 1, the troops engaged in constructing works, strengthening their positions and reconnaissance parties were sent out daily to observe the movements of the enemy in Atlanta and skirmishers were sent forward toward the city.

September 1, 1864, The Union Army 20th Corps was in position covering the crossing of the Chattahoochee River. The Second Division which included the 102nd NY Volunteers was at Pace's Ferry. A reconnaissance party sent out by the First Division and included some members of the 102nd NY found Atlanta was still occupied by the enemy. But that evening heavy explosions were heard coming from Atlanta, and a force was at once ordered from each division to make a reconnaissance.

September 2, 1864 Reconnaissance was sent from each division of the 20th Corps and finding Atlanta evacuated they took possession of it. The reconnaissance party from the Third Brigade under Colonel Coburn that entered the city was met by the mayor and the city was surrendered to them. On this and the following day, the whole corps, except the First Brigade of the Third Division, marched into the city and took possession of the works. The First Brigade left behind remained at the river to guard the railroad bridge until September 16, when it too was ordered up, leaving one regiment (the 105th Illinois) behind to guard the bridge. The 102nd NY Volunteers, as part of Third Brigade, was in the southwestern portion of the line from the East Point railroad to the McDonough Road.

September 3, 1864 The 20th Corps entered Atlanta at 11 a.m. The Confederates had left except for some calvary, and a few shots were exchanged with them. General Slocum wrote that the rebels marched on the McDonough road and destroyed 80 cars loaded with ammunition and also seven train engines. But the Union captured three engines, a few cars, 14 pieces of artillery, more than 2000 small arms and about 100 prisoners, as well as some stores. They quickly repaired the railroad. The siege of Atlanta lasted 42 days. Lieutenant Colonel Chatfield, who commanded the 102nd NY Volunteers reported, "During the whole campaign the officers and men behaved themselves with credit and bravery and most commendable patience, with a firm reliance that in the end success would crown their efforts and bring them near the time when the Stars and Stripes, under which they have so long and bravely fought, shall triumphantly wave over the whole of our once happy Union." WOTR Vol 38 Part II P. 292

On **September 4, 1864**, General Sherman issued Special Field Order No. 64. General Sherman announced to his troops that "The army having accomplished its undertaking in the complete reduction and occupation of Atlanta will occupy the place and the country near it until a new campaign is planned in concert with the other grand armies of the United States." Wikipedia: Atlanta campaign

The capture of Atlanta greatly helped the re-election of Abraham Lincoln in November and hastened the end of the war. That is an interesting story but not the purpose of this book. What is important to know is that General Hood led his defeated army away from Atlanta and to the west, opening the way for Sherman's March to the Sea and resulting in the virtual destruction of Hood's Army of Tennessee during the Franklin-

Nashville Campaign. In 1872, many of the fallen soldiers were relocated and reburied in the Patrick R. Cleburne Confederate Cemetery. For more information see: Wikipedia: The Patrick R. Cleburne Confederate Cemetary

September 10, 1864 Colonel David Ireland, who commanded the Third Brigade of the Second Division of the Union Army 20th Corps, which included the 102nd NY volunteers died from dysentery. General Geary of the Second Division would report that Colonel Ireland had commanded the Third Brigade for 10 months and had distinguished himself by gallantry in all the engagements in which his command participated, "In his death I lose a valued personal friend, and the country one of its noblest defenders." Of the Atlanta Campaign, General Geary reported, "Thus triumphantly has ended this campaign, unequaled in the present war for glorious victory over almost insurmountable difficulties and unsurpassed in modern history. Thus, has ended a campaign which shall stand forever a monument of the valor, the endurance, the patriotism of the American soldier. Four months of hard, constant labor under the hot sun of a southern summer; four months scarce a day of which has been passed out of the sound of the crash of musketry and the roar of artillery; two hundred miles traveled through a country in every mile of which nature and art seemed leagued for defense-mountains, rivers lines of works; a campaign in which every march was a fight, in which battles follow in such quick succession are so intimately connected by a constant series of skirmishes that the whole campaign seems but one grand battle which, crowned with grander victory, attest the skill and patience of the hero who matured its plans and directed their execution." WOTR Vol 38 Vol II P. 147

After they lost Atlanta, the Confederate army headed west into Tennessee and Alabama, attacking Union supply lines, as they went. Sherman was reluctant to chase Hood across the South, so he split his troops into two groups. Major General George Thomas took some 60,000 men to meet the Confederates in Nashville, while Sherman took the remaining 62,000 on an offensive march through Georgia to Savannah, tearing up everything they could as they went to destroy the will of the South to fight. History.com: Sherman's March to the Sea.

September 12, 1864 The 102nd NY Volunteers had been in the south part of Atlanta by the Macon Railroad but on this date they moved to another location. Lieutenant Colonel Chatfield reported, "the position of the brigade was changed to a better locality, nearer the city and about half a mile to the rear of the works, on which place another camp was laid out, this regiment being placed on the left of the brigade. While here the time was spent in drilling and preparing the men for an active campaign whenever called upon, and during a considerable portion of the time the regiment was employed in the construction of the new line of works then being built about the city." WOTR Vol 39 pt.1 p. 675

September 18, 1864 General Geary was busy writing his official reports about the Atlanta Campaign. The campaign took 123 days and during that time they fought ten battles, three times more skirmishes, and were under fire most of the time. He called it one grand battle and one grand victory from beginning to end. The Second Division suffered 2700 casualties, but the rebels had many more than that. General Geary's report would take 35 pages in the Official Records, which was five pages longer than General Sherman used to record the movements of his whole army.

Even though Atlanta was conquered, the Union troops remaining were vulnerable to Confederate General Hood's 30,000 men who could not take the city but could disrupt supply lines. General Sherman sent men to chase them, but the Confederates left a trail of destruction that caused Sherman to sever his supply lines and make his march to the sea.

Only the 20th Corps, which included General Geary's Second Division, remained to guard the city. They busied themselves with garrison duty and raiding for supplies.

September 25, 1864 The Union Army 20th Corps Second Division under General John Geary was reviewed and gave a good showing. FTCM p. 344 From the high ground near the review, Generals Geary and Williams could see where they were positioned for weeks outside the city facing this same high ground. The line looked less formidable from inside, but it still had a very strong row of obstacles and made the Generals realize it was a good thing they did not try a frontal assault. To protect his supply lines in his rear, General Sherman sent Newton's Division of the 4th Corps and Corse's Division of the 17th Corps to Chattanooga and Rome.

October 3, 1864 The October weather that year was rainy, cloudy and overcast. The rebels were causing disruptions in the Union supply lines and communications. The 4th Corps and part of the 14 Corps moved toward the Chattahoochee River to try and find where the railroad had been cut. For the 20th Corps in Atlanta, the mail and papers stopped a day before. General Sherman sent Major General Thomas with Morgan's Division of the 14th Corps to be in command in Tennessee.

October 8, 1864 The cool and rainy weather continued, and it was evident by now that General Hood had badly cut the railroad and the 20th Corps was without mail or news for 10 days. General Geary wrote about a new system of fortifications that was being built in Atlanta, the completion of which would destroy many of the finest buildings in the city. He also said that three divisions of Union troops were being sent back to Tennessee and Alabama to confront the rebels who were trying to cut the railroads behind them.

October 11, 1864 The 102nd NY Volunteer Regiment was sent on a foraging mission with General Geary. They left at 6 a.m. and marched in a southeasterly direction about 13 miles and bivouacked near South River about 8 p.m.

October 12, 1864 The 102nd NY Volunteers crossed South River as they were in charge of a portion of the Division supply train and went about 4 miles south of the river. Once there they filled their wagons with corn and returned to where they had been the night before.

October 13, 1864 The 102nd NY Left at 7 a.m. and again crossed South River. Two companies of the regiment were placed across a road leading in an easterly direction from the one traveled by the trains to guard against any approach by the enemy in that direction, under command of Capt. R. B. Hathaway. They marched about five miles south of the river with the remainder of the regiment, but then after a short time, they were ordered to move back to their previous camp-ground and guard fifty wagons filled with forage. That same day at 8 p.m., the command started their return and marched in the rear of the first 100 wagons of the train until about 3 a m on the morning of the 14th and bivouacked.

October 18, 1864 General Geary and most of the Second Division just returned from an extended foraging expedition. They went South of Atlanta toward Macon along the West Bank of the Ocmulgee River. The Division had two fights with the enemy and beat them both. They returned with lots of corn and sweet potatoes and continued to work on new fortifications for the defense of the city.

October 19, 1864 The October weather was still cool and rainy, and the 20th Corps still held the city of Atlanta alone. The rest of the men were dispatched to drive off Hood's Confederates and were successful. There were no other U.S. troops within 75 miles and the rebels had some troops in their front and skirmishing happened occasionally, enough to keep the men of the 20th Corps occupied.

October 24, 1864 The weather remained cool but was clear, or what General Geary would call in Pennsylvania an Indian Summer. The 20th Corps had devastated Atlanta for 20 miles around by foraging parties looking for food. News came to the men about the recent [primary] elections and the probable re-election of President Lincoln in November. The Southern Democrats under their command were upset by the news.

October 26, 1864 The 102nd NY Volunteer Regiment was again engaged in another expedition under General Geary. They left their camp at 6 a.m. and traveled along the Decatur road and once they reached Decatur they separated from the Third Brigade and formed an advance guard in front of their wagon train. They marched about 15 miles that day and bivouacked near Yellow River about 8 p.m. and then they performed picket duty for the division.

October 27, 1864 The 102nd NY Volunteers were camped near Yellow River on an expedition with General Geary. Colonel Chatfield reported, "On the 27th the picket was attacked by a small scouting party of the enemy's cavalry on both roads leading in an easterly direction, but their advance was checked by the force which I had thrown across these roads; remaining here until the evening of the 27th, when the column started about 8 p.m. and marched about seven miles toward Atlanta." WOTR Vol 39 p. 676

October 28, 1864 Somewhere on the outskirts of Atlanta, the Second Division was skirmishing with the enemy during a foraging mission and General Geary was reminded that it was the anniversary of his son Eddie's death. General Geary wrote that he was sad, but that God's will was done, and his name was blessed. As for the 102nd NY Volunteers, they returned to their camp in Atlanta as the foraging mission was over.

Moses Whitbeck was promoted to Sergeant on **November 1, 1864**. Details are limited, but the record shows Moses Whitbeck was returned to his unit sometime in autumn 1864. We can surmise that because he was promoted to Sergeant again that he was now back at his unit, about six months after receiving a serious head wound at the Battle of Dallas, was sent to Chattanooga and then Alexandria to recuperate. We can presume that he served in the campaign to Savannah.

November 5, 1864 The Second Division is getting ready to leave Atlanta and move upon Macon. General Geary wrote, "We are entirely cut off from the rest of our countrymen. This is not by compulsion, but by our own voluntary act. Atlanta will be destroyed and all the Rail Roads leading into it." APGTW P. 213

November 8, 1864 The 20th Corps is still in Atlanta. The men are in good spirits and ready to leave Atlanta.

November 9, 1864 The Second Division was attacked by Confederate Alfred Iverson's command at daylight in a surprise raid. They were repulsed after throwing over 100 shells at the Union forces with no casualties caused. Colonel Chatfield reported, "The enemy attacked the picket-line on the

Macon Road an advanced with a section of artillery and a few dismounted cavalry toward our works. The regiment was quickly moved into its position in the works and there remained awaiting any attack which the enemy might make. After shelling our line a short time the enemy retired. During the attack one man was slightly wounded by a shell. From this time until the commencement of the Georgia campaign the men were busy preparing for the active service which was soon to commence." WOTR Vol 39 p. 676

Savannah Campaign November 15, 1864

From **November 15 until December 21, 1864**, Union General William T. Sherman led some 60,000 soldiers on a 285-mile march from Atlanta to Savannah. One purpose of Sherman's March to the Sea was to scare Georgia's civilian population into abandoning the Confederate cause. Sherman's soldiers did not destroy any of the towns in their path, but they stole food and livestock and burned the houses and barns of people who tried to defend their property. The Union was "not only fighting hostile armies, but a hostile people," Sherman explained at the end of the campaign; as a result, they needed to "make old and young, rich and poor, feel the hard hand of war." Letter of William T. Sherman to Henry Halleck, December 24, 1864 The 20th Corps, which the 102nd NY Volunteers are part of were sent to Savannah.

November 15, 1864 The Union Army 20th Corps with the Second Division, led by General John Geary, broke camp in the early morning hours and at 7 a.m. moved out along the Decatur Road following the First Division. Shortly after passing beyond the line of old rebel works, they were forced to halt, on account of delays by the troops and trains in front of them and several hours elapsed before the road was sufficiently cleared for them to advance. The Second Division closely followed the First Division and then stopped in Decatur for dinner before marching close to Stone Mountain about 11 p.m. and camping for the night. General Geary reported, "The march during the day was continually delayed by halts and detentions. Caused by the miserable character of the animals in our train. The roads traveled were bad, the weather was beautiful. The distance marched during the day was fifteen miles." WOTR Vol 44 p. 269

November 16, 1864 The Union Army 20th Corps Second Division, which included the 102nd NY Volunteers, broke their camp near Stone Mountain at 8 a.m. and moved out in the front of the corps. They crossed Yellow River, at Rock Bridge, about 3 p.m. and went into camp about three miles beyond, having marched during the day 10 miles. General Geary reported, "The marching today was necessarily slow, owing to the bad character of the roads and bad condition of our animals. The country through which I passed was for the most part poor and undulating, and east of Yellow River the road crosses a number of swampy streams and steep ridges." WOTR Vol 44 p. 269-270

November 17, 1864 The Second Division of the 20th Corps, which included the 102nd NY Volunteers moved from camp again, this time at 5 a.m. and once again led the Corps and marched 17 miles on good roads through fine country and camped on the west bank of the Ulcofauhachee River.

November 18, 1864 The Second Division once again moved at 5 a.m. and led the 20th Corps and crossed the Ulcofauhachee River. They then struck the Georgia railroad at Social Circle, east of which they destroyed a considerable portion of the tracks and passed through Rutledge Station at noon. They stopped near there for dinner and destroyed the depot, the water-tank and other railroad buildings, and tore up and burned the train track. They had marched 18 miles and camped within two miles of Madison. General Geary reported, "The roads from Social Circle to Madison were excellent and the country was much superior to that previously passed through. Forage was abundant on every side, and during the day we made captures of horses and mules." WOTR Vol 44 p. 270

November 19, 1864 The Second Division left camp at 5 a.m. unencumbered by their wagons which would be brought along by the other divisions. They passed through Madison before daylight and went along the road that was parallel to the Georgia Railroad. They tore up some of the track and buildings as they went. They stayed the night at Buck Head Station but had to drive some Confederates off before they could. General Geary reported, "I sent on a detachment in advance of the main body to drive these scouts and whatever there might be of the enemy's calvary in the vicinity across the Oconee and to burn the railroad bridge across the river; also another detachment several miles above to destroy a large mill and the ferry-boats across the Appalachee. Both of these parties were successful. The railroad bridge, which was a fine structure, about 400 yards long and 60 feet high from the water, and was approached by several hundred yards of trestle work at each end, was thoroughly destroyed. At Blue Spring I halted and set my troops to work destroying railroad. Here at night encamped on the plantation of Col. Lee Jordan, on which I found 280 bales of cotton 50,000 bushels of corn stored for the rebel Government. All the cotton and most of the corn was destroyed. In addition to this my command destroyed elsewhere during the day 250 bales of cotton and several cotton gins and mills. I also destroyed in all to-day about five miles of railroad and a large quantity of railroad ties and string timbers." WOTR Vol 44 p. 270

The 102nd NY Volunteers tore up about 800 yards of track themselves, Colonel Chatfield reported, "started about 5 a.m.; marched about ten miles and bivouacked just beyond Buck Head and near the Appalachee River at 4 p.m. During the afternoon the regiment destroyed about 800 yards of railroad track on the Augusta railroad by tearing up the track and burning the ties." WOTR Vol 44 p. 313

November 20, 1864 The Second Division of the 20th Corps moved at 7 a.m. with rainy weather and roads that were very deep and swampy. Leaving the railroad, they moved toward the Oconee River which was reached two miles below the railroad bridge and then moved down parallel to the river to Park's Mill, which they burned. The bridge across the river at that place had been previously washed away and ferry boats were used at the crossing, but the Union Army destroyed them. They did receive some fire as they moved along the river from squads of rebel calvary on the opposite side. The rebels however, were soon driven off. General Geary sent a small party out from his command to cross the river near the burnt bridge and they went on foot seven miles to Greensborough, driving a small force of rebel calvary through the town and then taking possession of it. After remaining in undisturbed possession of the town for several hours, they convinced the inhabitants that most of General Sherman's army was close by with designs upon Augusta. This small party then returned safely to the Division, recrossing the river in canoes. General Geary reported, "I learned the next day that the enemy were tearing up the Georgia railroad at Union Point, seven miles east of Greensborough, apparently being possessed with the idea that General Sherman's army was moving on Augusta and using the railroad as it came. From all I could learn then and since, it is my opinion that my small command could, at that time have penetrated to Augusta without serious opposition." WOTR Vol 44 p.271

Leaving Park's Mill and having crossed Sugar Creek, they came to Glade's Cross-Roads where they took the road leading to the left. Moving one mile and a half on that road, they again turned to the left on a smaller road and camped at dark near the large tannery and shoe factory and store owned by James Denham. The tannery was one of the most extensive establishments of its kind in the South. Most of the leather stock and goods had already been carried off, A few boxes of shoes and leather were found hidden in a barn and were turned over to the quartermaster department for distribution. General Geary reported, "My skirmishers and foraging parties during the day's march spread through all the country between the Oconee and the route of march taken by the rest of the corps. A large number of splendid mules and beef cattle and some horses were captured, and the troops lived well on the produce of the country. Distance today, ten miles." WOTR Vol 44 p.271

November 21, 1864 A heavy rain fell the night before and continued through the day, creating very deep roads and swollen streams. After destroying entirely Denham's tannery and factory, they moved at 8 a.m. on the road to Philadelphia Church, then once there took the Milledgeville Road, crossed Crooked Creek and camped at where the road forked with one road leading to Dennis' Mill and station, the other to Waller's ferry, at the mouth of Little River. General Geary reported, "A very heavy, cold rain fell all day and it made their marching quite difficult. The country passed through was a rich one and supplies were abundant, distance marched, eight miles. The rain ceased toward night and the air became very cold. Among our captures today was Colonel White, of the thirty-seventh Tennessee Regiment. He had been in command of the post at Eatonton, and in attempting to escape from the other column of our troops fell into my hands." WOTR Vol 44 p. 270 The 102nd NY Volunteers marched about 10 miles through a drenching rainstorm and camped at Dr. Nesbitt's plantation.

November 22, 1864 The weather was extremely cold, but the men got moving at 6 a.m., taking the road to Dennis' Station. General Geary, having previously determined that it would be impossible for his command to cross Little River below the crossing of the railroad, because the locals had destroyed the ferryboats and because the bridge was burned, instead crossed Rooley Creek at Dennis' Station where they constructed a foot bridge for the infantry while the horses and wagons forded the stream. The Union forces burned a mill, a cotton gin, and a large amount of grain and cotton. Then they reached the railroad at Dennis' Station where they found the rear train of the other divisions just passing. They moved on behind the train to Little River, where they received orders to immediately go to Milledgeville. They crossed the river on the pontoon bridge, passing the trains with much difficulty, and reached Milledgeville at dark. Because other units got there first, they passed through the town then crossed the Oconee on the large bridge and went to camp on the left of the First Division, with their left to the river. They marched 20 miles that day and the nighttime weather was intensely cold.

November 23, 1864 While most of the 20th Corps Second Division stayed in camp, the Third Brigade, including the 102nd NY Volunteers, were sent to the Gordon and Milledgeville railroad where they remained until dark destroying railroad track. Colonel Chatfield of the 102nd NY Volunteers reported, "The regiment went with the rest of the brigade in the afternoon for the purpose of destroying the railroad running to Gordon on the Macon railroad; worked until dark and returned to the camp, this regiment having thoroughly destroyed about three-quarters of a mile of the track." WOTR Vol. 44 p. 313

Also that day, Union General Slocum's troops captured the city of Milledgeville and held a mock legislative session in the capitol building, jokingly voting Georgia back into the Union.

November 24, 1864 The Second Division of the Union Army 20th Corps moved out at 7 a.m. but because the road was completely blockaded with the trains the Division didn't really get into motion until 10 a.m.. The road they traveled that day was hilly, but in decent condition, so they marched 14 miles. Just before dark they crossed Town Creek over a bad bridge and set up camp near Gumm Creek.

November 25, 1864 The Division was on the road at 6:30 a.m. but once again due to the trains in front of them were not able to really move until 9 a.m. They marched 9 miles that day until they reached Buffalo Creek which was an extensive, heavy timbered, swampy stream being about ½ mile wide where the road passed through it. The stream or swamp there was divided into eight channels which were spanned by as many bridges, varying in length from 30 to 100 feet each. The bridges had been destroyed by the rebels, but reconstructed that afternoon under the supervision of Captain Poe, chief engineer on the staff of General Sherman. By dark, the road in front of the Second Division was open and they crossed and then camped a mile and a half east of the creek. The crossing in the extreme darkness of the night and through the swampy roads east of the creek was very difficult. During the night shots were exchanged between the pickets of the Second Division and some of Confederate General Joseph Wheeler's Calvary. The 102nd NY Volunteers camped just beyond Buffalo Creek.

November 26, 1864 The Second Division of the Union Army 20th Corps, which included the 102nd NY Volunteers, were up again early and moving at 6 a.m. They marched about two miles but once again were stopped by Wagon trains from other divisions for about two hours. Once they got going again, they marched about two miles until they caught up with the First Division under General Alpheus Williams who were skirmishing with Wheeler's Cavalry and driving them through Sandersville. The Second Division moved on to Sandersville, where they parked their wagon trains and left them under the charge of the Third Division. The division then moved on to Tennille, Station No. 13, on the central railroad. From there they moved eastward and destroyed two miles of the track and camped near a school-house four miles east of Tennille. One battalion of Michigan engineers under Major George Yates (who was later killed at the Battle of Little Bighorn), reported to General Geary and helped the division with the destruction of the rails and camped with the division that night. They marched 13-15 miles that day total. Colonel Chatfield reported, "Started at 5 a.m.; passed through Sandersville, Ga., and reached Station 13, on the Macon and Savannah Railroad, about 4 p.m., having marched fifteen miles in all. The regiment was ordered to go into position in advance of the brigade, so as to guard against any attack which might be made by the enemy's cavalry upon the troops who were at work destroying the railroad. This order was obeyed, and the regiment remained in line until after dark, when it bivouacked with the rest of the brigade near Station 13." WOTR Vol 44 p. 313

Also that day, in the Savannah campaign, In the late afternoon, elements of Union General Kilpatrick's 3rd Cavalry Division had reached the wooden railroad bridge north of Waynesboro, Georgia, and partially burned it before being driven off by troops dispatched from the Confederate Cavalry Corps of the Army of Tennessee by Joseph Wheeler.

November 27, 1864 The Second Division was on the march that morning at 7 a.m. and destroyed the railroad for four miles to a point where a road crossed the railroad 7 miles from Station number 13. From there they marched to Davisborough by a direct route and set up camp about 9 p.m. having marched 12 miles. Colonel Chatfield of the 102nd NY Volunteers reported, "broke camp at 6.30 a.m.; continued the destruction of the railroad until about 2 p.m., when our march was continued. Arrived at Davisborough about 10 p.m., having marched about twelve miles during the day. Placed the regiment on picket pursuant to instructions received from the division field officer of the day." WOTR Vol 44 p. 313

Meanwhile, nearby Sandersville suffered when General Sherman and his entire Left Wing marched into town early in the morning on November 26, but it almost was worse. His Left Wing had split after Milledgeville, but now it converged on two separate roads just outside of town. The night before, Confederate Gen. Joe Wheeler's cavalry had galloped into town with 13 prisoners, a dozen of which he placed in an improvised barrack in a store and one which was taken to the

Methodist parsonage and placed in the care of by the Rev. J.D. Anthony. Around midnight, an unknown vigilante mob subdued Wheeler's sentry, seized the prisoners, and took them out to a field, and shot them. Local citizens, fearing Sherman's wrath, quickly buried the bodies before sunrise. The next morning, Wheeler's cavalry briefly skirmished with the Union troops, as they approached the town, firing from the courthouse and other buildings, before saddling up and heading out in the opposite direction.

When Sherman reached town and learned of the skirmish and execution of Federal prisoners, he decided to burn the town of 500 to the ground. The Rev. J.D. Anthony pleaded with Sherman to spare the town, asserting that Confederate vagabonds killed the prisoners, and Wheeler's men — not the town — fired on Federal troops. The Rev. Anthony informed Sherman that only four men — three old and feeble — were left in town, the rest of the citizens were women and children. (Of Washington County's 1,460 eligible men, 1,502 signed up in 15 different Confederate companies…some signing up twice… perhaps the best record in the Georgia and maybe the South.) Something the Rev. Anthony said must have convinced Sherman, because he only burned the courthouse and downtown district, four cotton warehouses, and destroyed the train track. Worse, his men looted the county of all available food, creating hardship for the Georgians. But they were spared their homes. On Nov. 27, the Left Wing left town moving toward Louisville.

November 28, 1864 At dawn on November 28, Confederate General Wheeler suddenly attacked Union General Kilpatrick's camp south of Waynesboro and drove him southwest beyond Buckhead Creek toward Louisville.

Meanwhile, the Union 20th Corps marched east, destroying the railroad as they went. The Second Division demolished a section of track that the First Division was supposed to take care of according to General Geary. (An interesting side note here is that General Williams and General Geary often wrote contradictory reports about the same events.) General Geary detached Jones' brigade to guard the headquarters' trains to Station No. 11. With his other 2 brigades and the battalion of Michigan engineers, they destroyed part of the railroad from Davisborough westward that was about five miles and included a number of bridges, trestle-works and water tanks. While they were engaged in that work a portion of rebel cavalry in Ferguson's Confederate brigade fired at them for an hour and a half until driven off by skirmishers. General Geary reported, "The country through which the railroad passes from No. 13 to No. 11 requires description. It is a continuous morass, known as Williamson's Creek, or swamp. The stream is quite a large one, running in general direction parallel to the railroad and crossing it many times. The land in the vicinity of both sides is soft and swampy, with dense thickets of underbrush and vines. Through this swamp the railroad is constructed on an embankment of borrowed earth thrown up from the sides, averaging from six to ten feet in height. The superstructure consisted of cross-ties bedded in earth, with string timbers pinned to them upon which the iron rails were spiked. The mode of destruction was to tear up, pile, and burn the ties and string timbers, with the rails across, which when heated, were destroyed by twisting." WOTR Vol 44 p. 273

Shortly after dark, they returned to Davisborough and camped for the night, the distance traveled that day by the command was 15 miles. Colonel Chatfield of the 102nd reported, "moved at 7 a.m. back to point on the railroad distant some seven miles to continue the destruction of the railroad. Commenced tearing up the track bout noon; after working a short time a portion of the troops so engaged were fired upon by small party of the enemy. I ordered the flank companies to cease work and deploy as skirmishers on both sides of the railroad, to guard the remainder of the regiment against any attack and allow them to continue the destruction of the road. In short time thereafter, pursuant to orders received from Col. H. A. Barnum, commanding, I ceased work upon the railroad and marched back to join the main body of troops. This order I obeyed with much reluctance, as the destruction of the track was incomplete and many of the bridges, which were numerous along this portion of the railroad, were left undestroyed. As soon as the troops were drawn in we marched back to division and bivouacked at 8 p.m." WOTR Vol 44 p. 313

Also that day on the campaign, the Battle of Buck Head Creek (also known as Buckhead Creek and Reynolds' Plantation) was fought as the second battle of Sherman's March to the Sea. Union Army cavalry under Brig. Gen. Hugh Judson Kilpatrick repulsed an attack by the small Confederate cavalry corps under Maj. Gen. Joseph Wheeler but abandoned its attempt to destroy railroads and rescue Union prisoners of war.

November 29, 1864 The Second Division moved out at 6:30 a.m. and followed the main Louisville road for seven miles to Fleming's house, turning 90 degrees and then moved eight miles to Spiers Station (No. 11) which they reached at 1 p.m. After a short stop for lunch, they moved on following the road toward Station No. 10 and camped about 7 p.m. on the east side of a small creek. General Geary reported, "The roads traveled today were generally good and quite dry and hard west of Spiers Station. East of that place there was considerable swamp and marshy ground. The country through which we passed on the Louisville Road was excellent, the plantations being large and the buildings fine. After leaving that road the country is poorer and appears to be newly settled. Distance traveled was twenty-one miles." WOTR Vol 44 p. 273

November 30, 1864 The Second Division marched at 6 a.m. and went toward the First Division who had camped about a mile and a half away. When they got there, they discovered the First Division had not left yet. At 10:30, they followed that division north toward Louisville, leaving Jones' brigade again, and traveling about three and a half miles to a railroad bridge across the Ogeechee River. They had to repair the bridge first because the rebels had partly burned it, then they crossed it and then they destroyed that bridge and a wagon bridge and then continued toward Louisville. They traveled two miles then camped at dark on the east side of Big Creek, on a hill overlooking miles of the country, about two and a half miles south of Louisville. General Geary reported, "The country on both sides of the Ogeechee is an extensive swamp, with thick tangled growths. These swamps however have good sandy bottoms, and it was not difficult to pass through them. Distance marched was 10 miles. " WOTR Vol 44 p. 273

December 1, 1864 The Second Division was on the march at 7 a.m. in the lead position for the 20th Corps but behind the Ninth Illinois Mounted Infantry, following the road toward Millen. They crossed Big Dry Spring and Baker's creeks, passing through the camp of Carlin's Division of the 14th Corps, west of Baker's Creek and camped a mile and a half from Bark Camp Creek. General Geary reported, "The country passed through on this day's march was very swampy, although the roads in the main were very good. The facilities for forage were not as ample as on the previous days, the plantations being comparatively few, and although these few bore marks of having been cultivated, the stock and provisions had been mostly removed. The distance traveled was thirteen miles." WOTR Vol 44 p. 273-274

December 2, 1864 The Second Division was still in the lead so at 6 a.m. they moved and crossed Bark Camp Creek and went east in the direction of Buck Head Creek, and got there about noon. General Geary reported, "The roads traveled were excellent, following the course of a low dividing ridge. Passed but few plantations; among these was that of Doctor Jones, about 5 miles west of Buck Head Creek, one of the finest in this part of Georgia. Upon approaching the creek, I found a number of rail defenses, which had been erected a few days previous during a fight between the cavalry of Kilpatrick and Wheeler. The bridge was destroyed and the enemy's pickets fired upon us from the eastern bank. These were soon driven away by a regiment of my command, and the bridge was reconstructed by the Michigan engineers." WOTR Vol 44 p. 274 They crossed the creek in the afternoon about 3 p.m. and camped on the east side of the creek in the vicinity of Buck Head Church.

December 3, 1864 The Second Division having been assigned to the rear of the corps, were the last to leave at 11 a.m. and they followed closely behind the Third Division of the 20 Corps. The corps discovered that about five miles north of Millen, and not far from the railroad, there was a prison pen or stockade that until recently had confined about 3,000 union soldiers. The stockade was about 800 feet square, and enclosed nearly fifteen acres. It was made of heavy pine logs, rising from 12 to 15 feet above the ground. On top of the logs were mounted sentry posts about 80 yards apart. Inside of the stockade, running parallel to it at a distance from it of thirty feet, was a fence of light scantling, supported on short posts. This was the "dead line." About one third of the area, on the western side, was filled with earthen huts obviously made by prisoners. They found 3 dead soldiers there and gave them a proper burial. Through the eastern part of the pen ran a ravine with a stream of good water, but the atmosphere in the enclosure was foul and putrid. A short distance outside the stockade was a long trench with a board at the head that said "650 buried here." On higher ground, a short distance southeast of the stockade were two forts that had not yet been completed, and to the southwest there was another smaller one being built. General Geary reported, "This prison, if it can be designated as such, afforded convincing proofs that the worst accounts of the suffering of our prisoners at Andersonville, at Americus, and Millen were by no means exaggerated." WOTR Vol 44 p. 274 This prison would later become known as Camp Lawton. For the rest of the day the Division continued their march but were met by a long and almost impassable swamp and it took them until late in the night and early the next morning to catch up to the rest of the corps at Big Horse Creek. After having traveled a tough 10 miles, they briefly camped.

December 4, 1864 Even though they were last-in the night before and still in the rear of the Second Division of the 20th Corps, which included the 102nd NY Volunteers, they moved at 7:30 in the morning. About noon, they caught up with the

Third Division trains parked on the western side of Crooked Run. The eastern side of the stream was an extensive but level swampy tract of land across which trains of wagons could not travel until the road had been corduroyed, which was accomplished by the Michigan engineers. The wagon train from the Third Division did not leave until dark, and General Geary's command wasn't able to cross over until 11:30 p.m. They only traveled four miles that day and camped about a mile from Crooked Run. Colonel Chatfield reported, "continued the march until 4 a.m., having made about three miles, when the regiment bivouacked. At 7.30 a.m. again started; marched about six miles; the regiment was placed on picket duty for the night, with the One hundred and thirty-seventh New York Volunteers." WOTR Vol 44 p. 314

Also on the Savannah campaign that day was the Battle of Waynesboro. As Sherman's infantry marched southeast through Georgia, his cavalry under Judson Kilpatrick rode northeastward. After the numerically inferior Confederates withdrew, Kilpatrick entered Waynesboro the next day and destroyed a train of cars and much private property before being driven from the town by Wheeler.

Annoyed by Wheeler's constant harassment, Kilpatrick set out on the morning of **December 4** with his full division to attack Waynesboro and finally destroy Wheeler's small command. Early in the morning, Kilpatrick, now supported by two infantry brigades dispatched from Baird's Division of the XIV Corps, advanced from Thomas's Station six miles northward to burn the bridges over Brier Creek north and east of Waynesboro.

Finding Wheeler's Confederates deployed astride the road, Kilpatrick attacked, driving the Confederate skirmishers in front of them. The Union force then came up against a strong defensive line of barricades, which they eventually overran. As the Union advance continued, they encountered even more barricades that required additional time to overcome. After hard fighting, Wheeler's outnumbered force retired into Waynesboro and another line of barricades hastily erected in the town's streets. There, Wheeler ordered a charge by Texas and Tennessee troops in order to gain time to withdraw across Brier Creek and block the road to Augusta, which, at the time, appeared to be the objective of General Sherman's army. After furious fighting, the Union troops broke through and Wheeler's force hastily withdrew.

Finally reaching his objective of Brier Creek, Kilpatrick burned the rail and wagon bridges and withdrew. The supporting infantry brigades marched toward Jacksonboro and rejoined the rest of Baird's Division and camped at Alexander. They were followed that evening by Kilpatrick's command, which camped at Old Church on the old Quaker Road. Additional fighting over the next few days enabled Sherman to close in on Savannah.

December 5, 1864 The Second Division was on the move at 6:30 a.m. and crossed during the day Little Horse Creek, south fork of Little Ogeechee, and Little Ogeechee, destroying all the bridges after crossing them. Much of the route was through swamps which had to be corduroyed to get wagons through. At the south fork of the Little Ogeechee they destroyed a sawmill and marched about 12 miles that day.

December 6, 1864 The Second Division of the 20th Corps moved at 8 a.m., being the second division in the line of march. They had to stop twice because of slow movement upfront. After having moved about a mile, the division came upon the

Third Division parked waiting for a swamp to be corduroyed but few men working. So General Geary took some of his men from the Second Division and helped out while he supervised. They finished at dark, and the Third Division made their way over followed by the Second Division, and they both crossed a smaller swamp and then camped on the other side on good solid dry ground. General Geary reported, "The country was better than usual along the route today and foraging parties were quite successful. Weather warm and pleasant, Distance seven miles." WOTR Vol 44 p. 275

December 7, 1864 The morning was rainy, and the Second Division moved out at 7 a.m. but had to pass through a series of swamps so General Geary paired up the wagons with their regiments so that the men had to help the wagons through. This worked fairly well, and as they approached Turkey Creek the road improved. About 1 p.m., the rain stopped and the sun came out and the weather was warm. At 2 p.m., they reached Turkey Creek which was a wide and fordable stream with a good bottom. They had to wait for purposely felled timber by the rebels to be cleared. Then they crossed over and proceeded another three miles forward on an excellent road and camped within a half mile of Springfield having traveled 15 miles total on the day. Some Union soldiers encamped on the Eden Road after passing through Springfield. Others occupied Jerusalem Church, using its picket fence and hymnals for fires and engaging in skirmishes on the grounds.

December 8, 1864 The Second Division of the Union Army 20th Corps, which included the 102nd NY Volunteers and Sgt. Moses Whitbeck received orders to march in the front of the corps toward Monteith, leaving their wagons with the Third Division. They left at 6 a.m. on a road running southeast from Springfield. After going six miles, they took a small road branching off to the right, trying to find a road that led to Monteith. They followed this southwest road seven miles and camped in the woods about one and a half miles from the Louisville Road where the 17th Corps was moving. General Geary reported, "The looked for middle road was not found today. The roads were generally fair, although we crossed several small swamps. In them we found timber felled across the road. This was removed by our pioneers, without delay the march more than thirty minutes at any one time. Most of our route today was through pine forests. We passed a number of plantation houses in these forests and quite a large supply of potatoes, sugar cane, fodder, mutton, and poultry was obtained. It is worthy of note that the swamp water through this region is excellent for drinking purposes, being much superior to well water. Weather today pleasant, Distance thirteen miles." WOTR Vol 44 p. 276 The 102nd NY Volunteers camped that night at Wadley's Mill.

December 9, 1864 The 20th Corps Second Division moved out at 8:30 a.m. following the First Division. At Zion Church, they met the Louisville Road and there turned to the left on the main road running due east to Monteith Station. At Monteith swamp, five miles west of the station they met the best obstructions yet. The swamp was very large, about two miles wide where the road crossed it. Throughout the crossing, the rebels had chopped down great quantities of timber, and at the eastern end of the swamp had erected two small redoubts with flanking rifle pits. General Geary reported, "In these works they had two pieces of light artillery, supported by a small force of infantry. The artillery was so posted as to rake the road running through the swamp. While the division preceding me was engaged in movements for the dispersion or capture of the force opposing us, my command was halted and massed at the western side of the swamp. Receiving orders to that effect, I sent Jones' brigade rapidly forward to support Carmen's brigade of the First Division, which was working its way through to our right of the enemy's position. The services of this brigade were afterward found not to be required." WOTR Vol 44 p. 276 What he meant was that the First Division was able to overtake the rifle pits by themselves and did not need help. The First Division was commanded by General Alpheus Williams and the Second Division by General Geary. These two Division Commanders did not like each other and General Geary being the Politician from Pennsylvania was always making events appear better for his Division than may have been the actual case. They camped at dusk on good dry ground between two portions of Monteith swamp, having traveled six miles that day.

December 10, 1864 The order of 20th Corps march was First, Third and then Second Division last. The Second Division lead by General John Geary and containing the 102nd NY Volunteers and Sgt. Moses Whitbeck, had to guard all the wagons of the entire corps. They moved out last about 10 a.m. on the direct road to Monteith Station. The road was broad and solid and perfectly level. They passed the two redoubts at the east end of the swamp area that had been taken out the day before, and reached Monteith Station on the Charleston railroad, ten miles from Savannah, at noon. Much of the railroad track had already been destroyed by the preceding part of the Corps, so they stopped for lunch. Then afterwards, they moved toward Savannah on the Augusta Road, with the 14th Corps coming into Monteith right behind them. The advance part of the 20th Corps found the rebels behind fortifications about three miles from Savannah. The camped for the night near the Five Mile Post after having covered a distance of ten miles.

December 11, 1864 At 7 a.m., General Geary sent Barnum's brigade, which was the Third Brigade of the Second Division – and Moses Whitbeck and the 102nd NY Volunteers were a part of that – to perform reconnaissance between the Augusta Road and Savannah River. They found the enemy and their entire line was extended to the river. Skirmishers from the 102nd NY Volunteers drove a group of rebels from an advanced spot back to their main line, capturing a few prisoners. At 10 a.m., the Second Division was in line and established along an old rice field dike and had artillery backup. Most of the front line of the Second Division was covered in woods with the exception of their left which was in open ground within 250 yards of a large work on the riverbank in which the rebels had seven heavy guns. General Geary reported, "In front of my entire line were open fields affording a full view of the intrenchments held by the enemy. Immediately in front of these intrenchments were extensive rice fields flooded with water, and between the fields in my front and these flooded rice fields was a canal 25 feet wide and five or six feet deep, which was also filled with water. The sluice gates to these fields were all under control of the enemy, as was also the mouth of the canal, between which and my position was the large, advanced work before mentioned as being in front of my left. Besides this one the enemy had in my front three other works mounted with heavy guns, in their main line across the flooded rice fields. These guns all opened upon us, keeping up a steady fire throughout the day, but causing very few casualties. No reply was made by my artillery, but my skirmishers were advanced as far as possible and annoyed the enemy considerably." WOTR Vol 44 p. 276-277

Colonel Chatfield of the 102nd NY Volunteers reported, "At 7 a.m. moved out upon the main road to Savannah with the rest of the brigade; moved forward about three-quarters of a mile thereon and turned off to the left toward the Savannah river. In a short time thereafter the command hour I was ordered by Col. Barnum, commanding the brigade, to move forward the regiment and deploy it as skirmishers, connecting with the left of the One hundred and thirty-seventh New York Volunteers, which was already deployed and skirmishing with the enemy, and to continue my line until it reached the river, if possible. When but three companies upon the right had been deployed the river was reached. With the remainder of the regiment as a reserve I ordered the skirmishes forward. They had moved but about twenty-five yards when farther advance was prevented by a deep swamp, and the line had become so shortened as to render two companies sufficient to cover the space between the One hundred and thirty-seventh New York Veteran Volunteers and the river. I therefore ordered Company I to join the reserve. Moved the reserve forward nearly to the skirmish line and went forward with Col. Barnum to reconnoiter the position. To get over the swamp with any force it was found necessary to cross a narrow dike or road, which was commanded by the enemy's sharpshooter. At this time Capt. Maguire had succeeded in crossing with a few skirmishers, and he meeting with but little resistance, I ordered Company K, Capt. O. J. Spaulding, to cross the road, to quickly deploy, and with those already across advance toward the enemy's works. This was gallantly done, and the enemy driven into his main line of works. I immediately ordered the reserve forward, when the enemy opened upon the column with artillery, and the force being insufficient to carry the works by assault and unsupported, I ordered the regiment to form in line behind a natural dike, which had been previously occupied by the enemy and which was but 150 yards from their main line. During this attack the regiment had two wounded, Capt. Spaulding and one private, both slight. The regiment remained in this position and improved and strengthened the earth-work in their front sufficiently to protect it from the enemy's fire. At 11.30 p.m. it was announced to regimental commanders by Col. Barnum that a night attack was ordered and the plan detailed." WOTR Vol 44. p. 314-315

December 12, 1864 The 20th Corps strengthened their breastworks during the night so as to resist the rebel's heavy shelling. A steady artillery fire was kept up all day by the rebels, causing some Union casualties. The 102nd NY Volunteers were called to prepare to make an assault in the early morning hours, but the orders to attack never came. They stayed in place the whole day, dodging the enemy's sharpshooters and cannon shells. Apparently, according to Third Brigade commander Barnum, one of the supporting artillery batteries was delayed into getting into position that morning and also the First Brigade, which was supposed to be in front of the Second Brigade was accidentally marched the wrong way, causing all kinds of confusion and thus the delay. Because it was a cold night the waiting troops were pacing around and stomping on the ground to try to keep their feet warm and this alarmed the enemy, and the attack was called off. Colonel Chatfield reported, "At 12.30 a.m. the regiment was called up and preparations made to assault the enemy's lines at 1 a.m. At that hour the regiment was in readiness, but the attack was delayed and the regiment did not commence to move outside our line of works to get into position until about 4 a.m. This regiment was to form the left of the second line of the assaulting column, and the left wing had filed over and in front of our works, when the order for attack was countermanded, and I received orders to take my original position within our line of works, which I did. Remained here during the day, nothing occurring except being annoyed by the enemy's sharpshooters and few shells." WOTR Vol. 44 P 315

December 13, 1864 The artillery fire from the day before continued, but the rebels were getting better and starting to cause more damage. They had also put some sharpshooters in the upper story of a house near their fort on the riverbank. Colonel Chatfield of the 102nd NY Volunteers reported, "The enemy's sharpshooters kept up an annoying fire, and occasionally their artillery opened, with no other effect than the wounding of one man slightly." WOTR Vol 44 p. 315 Also, the previous night the rebels had captured some Union soldiers on Hutchinson's Island who had gone there to forage for supplies. General Geary then sent a larger force to the Island to hold the upper part of it. The Second Division also constructed a battery on the bank of the river near Jones' Brigade's position which was occupied by four 3-inch guns of Sloan's battery. These guns commanded the approaches up and down the river. General Geary reported, "The supplies of food and forage in our trains being mostly exhausted, our troops were now subsisting upon fresh beef, coffee and rice. Large quantities of the latter had been obtained upon the plantations in this vicinity, and a large rice mill on the Coleraine plantation, three miles up the river from my line, was kept constantly at at work. Forage for our animals was obtained from rice straw and from the canebrakes. There was also tolerable grazing in the woods. An advanced line of pits for my skirmishers and sharpshooters was constructed tonight in the open field, within plain sight of all parts of the enemy's line and within good musket range of it." WOTR Vol 44 p. 277-278 A Union division under William B. Hazen captured Fort McAllister, on the coastline south of Savannah. It allowed Sherman's men to link up with Navy ships holding supplies.

December 14, 1864 On the outskirts of Savannah, heavy and persistent artillery fire was kept up all day from the rebel batteries. The majority of their guns were 32-pounders, but they also used one 64-pounder and some light field pieces. General Geary received official orders announcing the capture of Fort McAllister, whose guns had been keeping a Union supply ship from reinforcing the troops. General Geary then sent a small wagon train to go get needed supplies. During the late morning, a rebel gunboat came up during the high tide in Back River, which is on the other side of Hutchinson's Island, and fired several shots into Jones' camp where the 102nd NY Volunteer regiment was and then withdrew. Colonel Chatfield of the 102nd NY Volunteers reported, "During the day the gun boat made its appearance in the river, nearly in a line with the left of my regiment, and opened fire upon the line with shot and shell from 6 ½ and 9 inch guns, from the effect of which I had five men slightly wounded." WOTR Vol 44 p. 315 The boat then retreated. The Brigade then readied themselves for the next day.

December 15, 1864 The rebels kept up their artillery fire, averaging over more than 300 shots per day. No reply was made by Union forces except by their sharpshooters, who were very active and accurate in their fire, causing much greater losses to the rebels than were produced among Union troops, according to General Geary's report. Geary's men were kept well concealed, and it was impossible for the rebels to make any correct estimate of the Union force. This day they also received for the first time since leaving Atlanta some New York newspapers dated December 10.

December 16, 1864 There was no change in the position or force the Second Division was in. The previous artillery fire from the rebels and the sharpshooting by the Union kept up. It was decided to place some heavy Union artillery on Geary's line, and construction was kept up during the night to accomplish that goal. The Confederates continued to send gunboats at high tide up the Back River to shell the Union position on Hutchinson's Island, but few casualties resulted. General Geary wrote, "As usual my command is the nearest to the city and in full view of it. The enemy have a strong line in our front, well defended with swamp dykes, rice-field marshes, etc., and are very defiant. We are therefore in the midst of the thunders of a siege." APGTW p. 216

December 17, 1865 The work on the Union fort continued even as the rebels did all they could to stop it by constantly trying to shell and shoot at it. During the late morning, the troops received their first mail in nearly six weeks, and there was much rejoicing. The 20th Corps First Brigade began construction of "Fort 2," to be a large lunette for heavy guns, but since it was in the open in front of Confederate lines, it had to be built quietly and at night. General Sherman demands that the city of Savannah surrender. He wrote, "I have already received guns that can cast heavy and destructive shot as far as the heart of your city; also, I have for some days held and controlled every avenue by which the people and garrison of Savannah can be supplied, and I am therefore justified in demanding the surrender of the city of Savannah, and its dependent forts, and shall wait a reasonable time for your answer, before opening with heavy ordnance. Should you entertain the proposition, I am prepared to grant liberal terms to the inhabitants and garrison; but should I be forced to resort to assault, or the slower and surer process of starvation, I shall then feel justified in resorting to the harshest measures, and shall make little effort to restrain my army—burning to avenge the national wrong which they attach to Savannah and other large cities which have been so prominent in dragging our country into civil war." William T. Sherman, Message to William J. Hardee, December 17, 1864, recorded in his memoirs.

December 18, 1864 The work on Fort No. 2 continued and that morning about 9 a.m. a heavy fog lifted. The Confederates tried to bombard the forts being built but had no success. The Union Army was building a third fort even closer to enemy lines than the first two. The city of Savannah refused Sherman's surrender demand.

December 19, 1864 All the generals commanding the different units of the 20th Corps had a morning meeting where they decided to storm the rebel works as soon as the heavy guns were in place and ready to fire. Fort No. 1 was finished that evening. The details from the First and Third Brigades continued work on the other two forts during the night under heavy artillery fire. Sloan's battery of 3-inch rifled guns had taken a position in a work thrown up to the right of Fort No. 3 and in an open field.

December 20, 1864 The artillery fire and sharpshooting of the past several days continued. By the evening, the Union men had constructed and made ready for use in the contemplated assault 200 large straw fascines to fill up ditches in front of the enemy works and also a large number of fascines made from bamboo cane. The canes were to be used for bridging the canal by laying them across "balks," which were furnished from the pontoon train for that purpose. The work on forts 2 and 3 was well advanced and would probably be completed that night. Three 30-pounder siege guns were installed in Fort 1. General Geary discovered that the rebels had built a pontoon bridge from Savannah over to the South Carolina shore and notified higher commands. That night, the movement of enemy troops was heard, and General Geary surmised they were evacuating. Confederate General Hardee decided not to surrender but to escape. He led his men across the Savannah River on a makeshift pontoon bridge. However, the rebels continued to fire artillery at the Union forts being built. Colonel Chatfield of the 102nd NY Volunteers reported, "About 9 p.m. my pickets on the left of the regiment reported that the men in the enemy's works in our front could be seen apparently moving to the right (their left) and soon thereafter the enemy could be heard crossing a pontoon bridge, apparently opposite the city. A strict watch was instituted and at about 10,30 p.m., becoming satisfied that they were leaving, went in person and reported these facts to the brigade commander. During this time the enemy kept up a vigorous fire from his artillery in our front." WOTR Vol 44 p. 277

December 21, 1864 After 3 a.m., the rebel artillery firing ceased. Second Division, Third Brigade pickets from the 102nd NY Volunteers advanced and found the rebels hastily retreating. Notice was sent to General Geary at once. General Geary notified Command HQ and then pushed his men forward in the direction of Savannah, hoping to overtake and capture a part of the enemy's forces. Skirmishers were deployed from the Third Brigade and swept over all the ground between the evacuated works of the Confederates and the Ogeechee canal from the river to the Augusta Road, while the main body of troops marched rapidly by the flank through McAlpin's plantation to the Augusta Road and on into the city. Just outside of the city limits, near the junction of the Louisville and Augusta roads, General Geary met the Mayor of Savannah and a delegation from the board of aldermen, bearing a flag of truce. From them, in the name of his commanding General, General Geary received the surrender of the city. He sent a messenger to command HQ that was delayed by Union troops questioning the messenger. In the meantime, the Second Division entered the city of Savannah at early dawn and before the sun was up raised their flags over the U.S. Customs house and county exchange. General Geary sent patrols out to make sure no soldiers or civilians were looting or burning and thus saved many millions of dollars' worth of cotton, ordinance, commissary stores, etc., which otherwise would have been destroyed. General Geary was made Commandante of the City, in honor of his capturing it.

The citizens of the town were now under protection of the flag that waved over them exactly four years since the state of South Carolina passed their secession act. Two regiments from Colonel Pardee's First Brigade, the 28th Pennsylvania and 29th Ohio Volunteers, were sent down to Fort Jackson early in the morning and gained possession of it and all the intermediate and surrounding works. The rebels burned all their gunboats. Many rebel flags were captured. General Geary wrote that his division alone captured 75 cannons with ammunition, 30,000 bales of cotton, 400 prisoners, and liberated many northern soldiers who had languished for months in the southern dungeons. General Geary wrote, "Oh how pleasant it is to bid the captive go free, none but those who do it can taste its ecstatic pleasures. God has vouchsafed me so many blessings that we will give all the glory to His great name." APGTW p. 219 General Geary also noted that during the campaign they had destroyed 26 and ½ miles of railroad track.

Colonel Barnum who was the commanding officer for the Third Brigade of the Second Division reported, "In justice to the officers and men of this brigade, it is here recorded that they were the first to discover the evacuation by the enemy of his works, the first to occupy them, the first to enter the city, the first to take possession of and guard all the captured ordinance and stores of every kind in and below the city and in the enemy's works from the Augusta Road to the river; that they captured the greater part of the prisoners taken, and…were the only organized body of our troops in the city." WOTR Vol 44 p. 310

In this action, the 102nd NY Volunteers led the way. Colonel Chatfield reported, "From 1 a.m. to 3 a.m. the sounds made by crossing could be so distinctly [heard], and every indication of the evacuation of the city becoming so apparent, an advance was ordered by Col. Barnum, who had come up in person to my position, to be made by ten men from my regiment, to reconnoiter the position in our front and discover whether or not the enemy was there. In short time they reported the line evacuated, and at 3.20 a.m. I entered the first line of the enemy's works with the regiment, finding seven guns in position and a large quantity of ammunition, &c., destroyed. In a short time the men sent forward reported the enemy's second line across the canal also evacuated. In obedience to orders from the brigade commander, I detached one company to guard the guns captured, and with the line where we halted and awaited the coming up of the remainder of the brigade. Detached two companies to take possession of and guard the guns in this line from the Augusta Road to the river. At 4.15 a.m. an advance toward the city was ordered. My regiment leading, marched rapidly forward until we reached the Augusta Road, when I ordered one company of the column as skirmishers; moved forward very rapidly and with no opposition except a few shots fired upon the advance guard from the bridges crossing the canal, and entered the city at daylight, capturing some few stragglers from the enemy and a large amount of stores of all kinds." WOTR Vol 44 p. 315

From **December 21 until January 19, 1865**, the 20th Corps was in Savannah performing garrison duty. General Geary became commander of the city of Savannah.

December 22, 1864 General Sherman sent his famous telegraph to President Lincoln announcing the news of the capture of Savannah, "I beg to present you as a Christmas gift the city of Savannah with 150 heavy guns & plenty of ammunition & also about 25000 bales of cotton." Telegram from Sherman to Lincoln. President Lincoln was delighted to hear the news. Responding the following day, he wrote, "My dear General Sherman. Many, many, thanks for your Christmas-gift – the capture of Savannah."

Colonel Chatfield of the 102nd NY reported, "During the whole of the campaign both the officers and men of my command have behaved well, and it is sufficient to say have done their whole duty as becomes veterans. It is difficult, where all have so will performed their part, to make any distinction, yet I would especially mention Capt. O. J.

Spaulding, commanding Company K; Capt. H. M. Maguire, Company C, and First Lieut. T. W. Root, acting adjutant, as deserving special mention for their bravery and vigilance throughout the campaign." WOTR Vol 44 p. 316

The Savannah Campaign was over, and now the Army would be tasked with protecting the town and the citizens. Colonel Chatfield reported, "I respectfully submit the following report of the operations of this regiment during the occupation of the city of Savannah, Ga., and the campaign through South and North Carolina ending at this place, On the 22d of December, 1864, pursuant to orders received from Col. H. A. Barnum, commanding brigade and provost-marshal of the Western District of Savannah, I was assigned to the command of Sub-District No. 1, and ordered to preserve order and cleanliness through out the same with this regiment. In pursuance of said orders guards and patrols were established and safeguards furnished where necessary, good order established and kept, and care taken that the rights of the private citizens should be kept inviolate. The regiment continued on this duty until 19th day of January, 1865." WOTR Vol 47 p. 760

1865

January 19, 1865 The Union Army 20th Corps, which included the 102nd NY Volunteers, was in Savannah, Georgia, from the end of the Savannah campaign on **December 21, 1864**, on garrison duty. This meant that they were in charge of District No. 1 to preserve order and keep it clean, and thus they made patrols and certain safeguards to make sure the public was protected. On this day, the Second Division was relieved by Union General Grover's Division of the 19th Corps, specifically the 14th New Hampshire Volunteers. As many men as possible had been given new uniforms and clothing. The other Divisions of the 20th Corps already had left and crossed the river and advanced on the other side of the river toward Purysburg. The Second Division was supposed to catch up, but there had been considerable rain, so the roads were unpassable for troops or wagons. The Division was therefore given instructions to await further orders. WOTR Vol. 47 p. 760

Colonel Chatfield reported, "The regiment citizens on this duty until 19th day of January, 1865, when myself and command were relieved by the Fourteenth Regt. New Hampshire Volunteers. The command having been put in readiness and supplied with clothing, as far as it was possible to obtain the same, pursuant to orders received from the general commanding brigade, on the 27th day of January, 1865, it moved out of the city upon the Augusta Road, marching to Sister's Ferry via Springfield, in charge of a large portion of the corps train, arriving at Sister's Ferry on the 29th day of January 1865." WOTR Vol. 47 p. 761

Campaign of the Carolinas January to April 1865

The 102nd NY Veteran Volunteer regiment was led by

Lieutenant Harvey S. Chatfield and Major Oscar J. Spaulding and was part of 20th Corps Army of Georgia. Moses Whitbeck was a sergeant in the 102nd, Company G, having survived a head wound at the Battle of Dallas in 1864. After spending a month in Savannah, Union General William Tecumseh Sherman headed his army north. This action would be called the Carolina Campaign. The 102nd NY was in the Third Brigade of the Second Division, and was marching on the left flank of the Union armies. To confuse the Confederates, Sherman sent his Left Wing westward toward Augusta and his Right Wing eastward toward Charleston. After several days, however, Sherman brought the wings together and advanced north toward Columbia.

In the Carolina Campaign, The Yankee soldiers took particular delight in carrying the war to South Carolina, the symbol of the rebellion. It was the first state to secede and the site of Fort Sumter, where South Carolinians fired on the Federal garrison to start the war in April 1861. The men were instructed to protect the public and their property as much as possible, but if people fired at Union forces from homes and buildings, then they were not protected anymore. These orders were largely ignored and there were "bummers" who put flame to as much as they could. Excerpt taken from History.com: Union army sacks Columbia, South Carolina.

The Second Division under General Geary was part of the Left Wing, which traveled through Blackville, Lexington, Winnsboro, Fayetteville N.C. and then on to Averasborough and Bentonville before the Confederate surrender at Durham Station. They began the campaign with slightly more than 5000 soldiers and 261 officers, 159 wagons, 33 ambulances, and more than 500,000 rounds of ammunition. Each brigade had a pioneer company of about 30 men and a tool wagon loaded with axes, picks and shovels.

January 24, 1865 The rains that had been hampering movement since January 19th finally stopped and the Second Division could start to prepare to move. The Third Brigade that the 102nd New York Veteran Volunteers were in was led by Brigadier General Henry. A. Barnum and the 102nd was led by Lieutenant Colonel Harvey. S. Chatfield.

January 27, 1865 The Second Division marched and left Savannah via the Augusta Turnpike. General Geary reported, "The morning was bitterly cold and the roads were frozen hard; these thawed a little during the day and the ground broke up, rendering their condition very bad; encamped at 3pm twelve miles from the city; my trains got into camp by dark." WOTR Vol 47 p. 682 The 102nd NY Veteran Volunteers were in charge of a large part of the Corps' wagon trains, as they were in the Third Brigade, which was in the back of the division. They crossed the Saint Augustine Creek and bivouacked near Montieth having marched 12 miles that day.

January 28, 1865 The Second Division moved out at 6 a.m. on the Augusta Road and marched four miles, then turned to their left on the McCall Road and marched toward Springfield and camped at Widow Bird plantation. It was very cold and thawed a little in the middle of the day and that caused the roads to be bad enough to need to be corduroyed, but they made 14 miles that day. The Third Brigade was camped near Little Ebenezer Creek.

January 29, 1865 The Division moved out at 6:30 a.m., marched through Springfield, forded Jack's Creek, and then about a mile and a half past turned to their right on the Sister's Ferry Road and then crossed a bad swamp at Ebenezer Creek.

The swamp required a lot of corduroying to get past. They camped on the plantation of Judge Mallette three miles from Sister's Ferry and made 12 miles that day.

January 30, 1865 A pontoon bridge was laid down at the Ferry, but the other side of the Savannah River was swampy for 2 miles with the road under water sometimes as much as 12 feet. But the waters were receding, and the Division with the help of pioneers and axe men were working to corduroy the roads enough to get through. The 102nd NY Veteran Volunteers supplied some of the men.

January 31, 1865 The Second Division remained in camp while heavy details worked on the roads. The rebels had planted some torpedoes and flooded the area. General Geary reported, "Heavy details at work on the road across the river; a number of torpedoes were found embedded in the road; some of these exploded; all others that could be found were carefully removed. The work on the road was one of the greatest difficulty on account of the depth of the water in the sloughs and over the surface of the country generally and of danger on account to the torpedoes; weather warm and clear; water in the Savannah River gradually falling." WOTR Vol 47 p. 682 The 102nd NY provided a small detail to help with the removal of the torpedoes.

On February 3, 1865 A unit of 1,200 Confederate soldiers, commanded by Major General Lafayette McLaws, attempted to prevent Slocum's wing from crossing the Salkehatchie River at the Battle of Rivers Bridge. The Federals flanked their opponents by crossing downstream, forcing McLaws to concede the crossing. Nearly unopposed for the next two weeks, Sherman's soldiers constructed bridges and corduroy roads that enabled them to traverse the rugged terrain. As they moved north, they cut railroads and laid hard hands on South Carolinians in their path.

As for the Second Division, they were still working to repair the roads for the last few days and have not been able to move. General Barnum of the Third Brigade reported, "remained in camp until February 4, doing picket duty and furnishing details to build corduroy roads. Dress parade was held each day by brigade during our stay here. A thorough inspection was also made by the brevet brigadier-general commanding." WOTR Vol 47 pt. 1 p. 750

February 4, 1865 General Geary moved the Second Division to the river so they could cross when it was their turn. He reported, "By 10 a.m. the calvary train was out of my way and my command began crossing. The rain ceased about 9am; the roads were in terrible condition. Crossed without serious delay and filled my supply train and the haversacks of the men at the temporary depot established at the upper landing, two miles above the pontoon bridge." WOTR Vol 47 p. 682 General Geary was ordered to take Selfridge's Brigade from the First Division, who had been guarding the depot and also General Kilpatrick's calvary train of 250 wagons with him and his command. They moved toward Robertsville and found the road through Black Swamp utterly impassable for three miles. General Geary left Barnum's Brigade, which included the 102nd NY Volunteers with the trains at the swamp while heavy details worked during the night to corduroy the road. The advance part of the Division that crossed over made nine miles that day. General Barnum reported, "halted at a landing two miles above Sister's Ferry and drew nine days rations of bread, coffee, and sugar, and four of meat, the men carrying three days' rations and the balance being taken in the wagons; moved on to a large swamp and went into camp about 6 p.m. Day's march eight miles and a quarter." WOTR Vol 47 p. 750

Colonel Chatfield reported, "Here the command remained until February 4, 1865, awaiting the completion of a road across the swamp on the Carolina side of the Savannah River, and furnishing a small detail to assist in removing torpedoes form the road and corduroying the same. On the morning of the 4th of February, 1865, the regiment moved out at daylight, crossed the river and proceeded on the march toward Blackville, passing through Robertsville and Lawtonville, and crossing the Coosawhatchie and Salkehatchie Rivers, the latter at Buford's Bridge, and reaching Blackville on the Charleston and Augusta Railroad." WOTR Vol 47 pt. 1 p. 761

February 5, 1865 General Geary put 1500 men to work on the road through the swamp and by noon the head of the wagon train had crossed. At 1 p.m., the Second Division moved across and then camped at a crossroad near Trowell's farm, eight miles from Robertsville. General Geary reported, "Near Mr. Trowell's house we found three soldiers of our army, who according to the testimony of negroes, had been pointed out by Mr. Trowell to some of Wheeler's calvary and by them shot in cold blood. Their bodies were found in the bushes not far from the house, where they were thrown by the murderers. I had them buried and Trowell's house and other property destroyed, and he was taken with us to be tried as accessory to the murder. Roads to-day after leaving the Black Swamp were quite good; weather clear and warm; distance nine miles." WOTR Vol 47 p. 682 The 102 NY Volunteers as part of Third Brigade camped near Steep Bottom.

February 6, 1865 The Second Division moved out at 6 a.m., taking the road to Lawtonville, passed through Lawtonville, then took the road toward Beech Branch and camped near Mears' store. General Geary reported, "The roads to-day were bad; weather warm. Toward evening it began to rain. The country passed through yesterday and to-day had been quite a rich one. The planters had fled to the upper country and the plantations now looked desolate. Most of the supplies had been carried off by the divisions preceding me." WOTR Vol 47 p. 683 The Third Brigade camped near Beech Branch Post Office.

February 7, 1865 There was heavy rain the previous night and all that day. The roads were in very bad condition and had to be corduroyed. At noon they reached the Coosawhatchie Swamp and found that it was swollen because of the rain and was three and a half feet deep and 300 yards wide with a treacherous bottom. Since there was no bridge across, 600 pioneers and axe men were set to work constructing a foot bridge and corduroying the entrance to the water. By 4 p.m., the Division began to cross. General Geary reported, "The bottom of the stream worked into deep holes of a quicksand nature, so that it was necessary frequently during the night to halt the trains, send the pioneers waist-deep into the stream and construct corduroy road three or four feet under water, pinning it down to prevent it from floating. In this way about one-half of the train crossed during the night, which was dark and rainy, thus adding to the discomfort of the occasion. But I had received orders from Major General Slocum commanding Left Wing, who was with my Column, that we must push forward as rapidly as possible, and no effort was spared. Portions of my command were, therefore, kept all night at work getting the trains across in the manner described; distance six miles and a half." WOTR Vol 47 p. 683

February 8, 1865 The Second Division moved out at 6 a.m.

and marched toward Buford's Bridge on Big Salkehatchie River. The weather was clear but very cold. The roads had been repaired by preceding troops and they crossed Jackson's Branch at the Augusta and Pocotaligo Railroad before crossing the Big Salkehatchie at Buford's bridge. General Geary reported, "This crossing is a succession of twenty-three small bridges, with intervening causeways, in all over a half-mile long. The stream is wide, deep and swampy. On the northern side of it was a strong line of works, with four embrasures, which command the bridge or causeway so completely that any direct attack against a force holding those works would have been useless. Encamped three brigades and most of my trains on the north side of the stream. Late in the night the crossing became very bad, and Selfridge's brigade, with fifty wagons, remained on the south side; distance 14 miles." WOTR Vol 47 p. 683-684

February 9, 1865 The Second Division moved at 6 a.m. and took the road to their left toward Blackville. General Geary reported, "Found the roads to-day generally good, passing through a well-cultivated rolling country. For the first time in this campaign my foragers found an abundance of forage and supplies; some of them went as far as Barnwell, and all returned well laden. Encamped at 3pm within a mile of Blackville…weather all day cold and freezing; a little snow fell; distance 18 miles." WOTR Vol 47 p. 684 General Barnum of the Third Brigade reported that the men bivouacked that night, which meant their wagons with the tents did not catch up to them.

February 10, 1865 General Geary marched his division to Blackville then to Duncan's and across the river there. They crossed before dark and camped on the plantation of Mr.

Winningham. General Geary reported, "Duncan's Bridge (better known among the inhabitants as New Bridge) comprises six bridges with causeways connecting them, the entire crossing being about one mile in length. Three of these bridges, including those across the two main channels of the South Edisto, had been burned by the enemy, and required much work to repair them. The country across the Edisto is a rich one, and the resources for subsistence and forage were abundant. Distance ten miles." WOTR Vol 47 p. 684 The Third Brigade went to the South Edisto River and crossed it at the upper end of Fair's Island and bivouacked by the river.

February 11, 1865 The Second Division stayed in camp as 2,000 men were at work repairing the bridges and corduroying the causeways. General Barnum reported that heavy details of men were sent to corduroy the road over the swamp. WOTR Vol. 47 p. 751

February 12, 1865 The Second Division marched at 7 a.m. and was in the lead of the Corps, as they marched toward Columbia via Jeffcoat's Bridge. General Geary reported, "Near the crossing of the Ninety-six road we met a small force of the enemy's calvary and exchanged shots with them. On reaching Jeffcoat's Bridge we found it burned, and the enemy holding the north bank of the North Edisto. The only approach to the bridge, except on the road, was through a swamp, covered with dense tangled growth of bushes, vines and briers. I deployed skirmishers on each flank, from the 5th Ohio Veteran Volunteers and 147th Pennsylvania Veteran Volunteers, who made their way with great difficulty through these swampy thickets, and drove the enemy from the opposite bank. The main channel here was very deep and the bridge of heavy timbers was effectually destroyed. On the opposite side was another extensive swamp, through which the road was built in the form of a causeway. The farther end of this causeway the enemy held and from their position swept the road and bridge with discharges of shell and canister from two pieces of artillery. On each side of the causeway the swamp was too deep to be waded. My troops held both ends of the bridge and a small earth work was thrown up. At dark the firing ceased and the battalion of First Michigan engineers, which had been obliged to cease working at the bridge during the afternoon, on account of the raking fire, resumed work, The weather to-day was clear and cold; roads were good; distance, twelve miles." WOTR Vol 47 p. 684-685

The Third Brigade, which included the 102nd NY Veteran Volunteers and Sgt. Moses Whitbeck, camped that night near Jeffcoat's Bridge on the North Edisto River.

February 13, 1865 The bridge was repaired by 1 a.m., and General Geary immediately sent skirmishers forward and discovered that the rebels had vacated the area. General Geary reported, "My skirmishers met those of the enemy intrenched at a bridge across a mill stream three-quarters of a mile from the river, and after a sharp encounter drove them and captured their works. At a fork of the road just beyond the enemy attempted to stand behind rail barricades, but were quickly driven from them." WOTR Vol 47 p. 685

The Third Brigade moved across the North Edisto River at 5 a.m. After the rebels showed a thin skirmish line, the 60th NY was pushed forward as skirmishers to drive them off and were then relieved by the Third Division of the corps who passed them. The brigade continued their march at dark and went until 9 p.m. then camped having only traveled five miles.

February 14, 1865 The Second Division was in the middle of the corps now behind the First Division, and they left at 8 a.m. General Geary reported, "Our route was by the direct Columbia road to its intersection by the Orangeburg and Lexington roads, where we turned to the left toward Lexington and encamped at the intersection of this road with that leading from Columbia to Augusta via Horsey's Bridge. Weather cold and rainy; roads good; soil sandy and poor; country settled by poor whites; forage scanty; distance seven miles." WOTR Vol. 47 p. 685

February 15, 1865 The Second Division was leading the corps and were disencumbered and following the Lexington Road. When they got close to Congaree Creek, they skirmished with rebel calvary who had just burned the bridge there and were guarding the crossing behind some log breastworks they had built at the end of the bridge. General Geary reported, "The sides of the creek were swampy, with dense thickets, and the stream was four or five feet deep. My skirmishers penetrated the thicket to the stream; a few of them waded it, and while they gained the enemy's rear another portion charged directly on the bridge, which was thus gained without any loss on our part. It was quickly rebuilt with poles and rails, and we marched rapidly forward. Lively skirmishing with the rebel calvary continued during the remainder of the day, my skirmishers meeting them at every ravine and hill and driving them on the run so rapidly that the main column could not keep up." WOTR Vol 47 p. 685 When they reached Red Bank Creek, the rebels tried to destroy the bridge but didn't have enough time. When they reached the Two-Notch Road, they again skirmished with rebel calvary and drove them off toward Columbia. The corps then set up camp there.

The 102nd NY Veteran Volunteers, as part of the Third Brigade, marched at 7 a.m. that morning and marched until they reached about two miles from Lexington and started setting up camp at 3 p.m. near Two-Notch Road. Then, about a half hour later, they received orders from General Geary to move forward and occupy the town. When they got about a mile from the town, they sent the 137th NY Regiment forward as skirmishers with the rest of the brigade behind in support. They moved forward to the town and through it, pushing a large rebel calvary group out. The brigade then occupied the town, protecting the citizens and property. General Geary reported, "A large force of the enemy's calvary was reported in sight in and around Lexington. On reaching a hill over looking the town and within easy artillery range of it, I posted my artillery and advanced the skirmish line. The rebel calvary retired before us and the town was occupied and held by Barnum's brigade without any opposition... Private property was strictly protected while my troops occupied it and no houses were burned." WOTR Vol 47 p. 686

Colonel Chatfield reported, "On the 15th of February, 1865, reached Lexington Court-House, which was ordered to be occupied that evening by Brevet Brig.-Gen. Barnum, commanding the brigade, in pursuance of which the town was entered without any serious opposition. This regiment was thrown across the Columbia road and, in compliance with directions received from general commanding brigade, care |was| taken that no buildings be burned or citizens molested in the exercise of their usual avocation." WOTR Vol 47 p. 761

February 16, 1865 Union General Sherman was able to concentrate his army faster than Confederate General Beauregard had anticipated and arrived at Columbia on the afternoon of February 16. The only Confederates defending the city were small detachments from Maj. Gen. Joseph Wheeler's cavalry corps, Maj. Gen. Matthew Butler's cavalry division, and Lt. Gen. Stephen D. Lee's corps from the Army of Tennessee. Heavily outnumbered, Lt. Gen. Wade Hampton ordered an evacuation of the city without a fight, although they destroyed all bridges over the Congaree and Saluda Rivers in an attempt to delay the Union forces. When Confederate General Wade Hampton's cavalry evacuated Columbia, the capital was open to Sherman's men.

As for the Second Division, they were in the rear of the corps guarding the trains of the corps and there did not move out until 10 a.m. They reached within about four miles of Columbia and did not take action that day.

Sherman occupied Columbia on **February 17, 1865**. The Second Division was behind the First Division, so they did not march until 9 a.m. They went toward Leaphart's Mill on Twelve Mile Creek. There they met the 14th Corps that was heading out toward the Saluda River. General Geary reported, "There we found the Fourteenth Corps marching toward the Saluda River, and encamped with the rest of our corps, while the Fourteenth Corps, Kilpatrick's calvary, and our wagon trains were to push forward and cross the Saluda at Hart's Ferry during the night, if possible. Two corps of the rebel Army of the Tennessee (Cheatham's and S.D. Lee's) were reported to be to-day beyond Lexington, moving across the Saluda River; the aggregate force of the two corps is estimated at from 5,000 to 8,000; distance, five miles." WOTR Vol 47 p. 686

February 18, 1865 The Second Division was detained that morning by the wagons of the groups ahead of them and did not cross the pontoon bridges at Hart's Ferry until 11 a.m.

They followed the 14th Corps toward Freshley's Mill, which is on the Broad River near the mouth of Wateree Creek. General Geary reported, "encamped at 4 p.m. at Ravencroft's Mill, the wagon train of the Fourteenth Corps being parked a mile ahead; weather delightful; roads generally good; country very hilly and well farmed; north of the Saluda the soil changes to a slaty clay, with quantities of silex and occasional granite bowlders; it is well watered with running streams; distance, eight miles." WOTR Vol 47 p. 686 Third Brigade marched at 8 a.m. and was in charge of the trains. They crossed the Saluda and made it to Metts' Steam Mill where they bivouacked for the night.

February 19, 1865 The Second Division was now in the rear of the corps and therefore did not march until 2 p.m. They followed the Alston Road two miles and then turned onto the road leading to Freshley's Mill and sent Mindil's Brigade one and a half miles up the Alston Road to hold it. About a half mile later the Division came up to the trains of the First Division. So they waited several hours for the traffic ahead of them to clear and then they took the main road to Freshley's Mill. Here they camped to the rear and to the right of the First Division with their right resting on the river near the mill. The Third Brigade, which contained the 102 NY Volunteers and Sgt. Moses Whitbeck, did not march until 3 p.m. and did not reach their spot until 9 p.m.

February 20, 1865 General Sherman left Columbia on this day. As he did, much of Columbia burned to the ground in a fire that has engendered controversy ever since. As the best evidence tells us, the destruction of Columbia was a tragic accident. Retreating Confederates set cotton on fire, and the burning embers were carried by the wind.

Some cotton bales continued to smolder during the day, and the high winds whipped them into a blaze as well that evening, spreading more embers around. Some Union soldiers, drunk on the liquor provided them by well-meaning but mistaken civilians, set fires themselves, but the record shows that more Union soldiers tried to stop the fires but were unable to do so. More than two-thirds of the city was destroyed. Already choked with refugees from the path of Sherman's army, Columbia's situation became even more desperate when Sherman's army destroyed the remaining public buildings before marching out of Columbia. Dead Confederates.com: The-burning-of-columbia-february-17-1865

Union Brigadier General Alpheus Williams wrote in one of his letters, "South Carolina will not soon forget us. A blackened swath seventy miles wide marks the path over which we traveled…the first gun on Sumter was well avenged."

The departing Union Army left South Carolina's houses "comfortless and shabby, and its people at home rusty, ignorant and forlorn." In language foreign to his letters that he wrote to his daughters in 1862 and early 1863, Williams decided of South Carolinians that "the tornado of war may do them good in the end." Excerpts taken from: historynet.com:blue-gray-war-soldiers-letters

As for the 20th Corps Second Division, they were not involved. They were in the center of the corps and did not march until 2 p.m. Following the First Division, they crossed over Broad River on a long pontoon bridge at Freshley's Mill and moved toward Winnsborough. They crossed the Abbeville railroad and forded Little River. General Geary reported, "A short distance from the river we crossed the Abbeysville railroad, which is a cheap structure of stringer track and strap rail. Following a very miry and unfrequented road through woods and fields, we forded Little River, a deep, rapid stream thirty yards in width, and at Colonel Gibson's house entered a main road to Winnsborough. Here, turning to our left, we moved forward on this road, which we found an excellent one, through very hilly country, and encamped within nine miles of Winnsborough. The country on our route to-day was a rich one, and forage and supplies were plentiful." WOTR Vol 47 p. 687 The Third Brigade bivouacked at a cross roads near Kincaid's house having traveled nine miles that day.

February 21, 1865 The Second Division was unencumbered and moved out early toward Winnsborough, reaching it unopposed about 11 a.m. General Geary reported, "When with-in two miles of the town I saw heavy smoke arising from it, and double-quicked my two advance regiments in order to reach it in time to arrest the conflagration, this we effected with much labor, my troops performing the part of firemen with great efficiency. About one square was burned before the fire could be arrested. A large number of foragers from various corps were found in the town. These were sent to their commands, and Brevet Brigadier General Pardee, with his brigade, was directed to occupy the town, while my two other brigades commenced destroying the railroad northward, three miles and a half of which they destroyed most effectually during the afternoon, burning the ties and other timbers and twisting every rail. Winnsborough is a pretty town of about 2,500 population, the seat of justice for Fairfield District. Among its residents were many refugees from Charleston. The surrounding country is well farmed and furnished abundance of supplies, which were brought in by our foraging parties; distance to-day, nine miles." WOTH Vol 47 p. 687

The Third Brigade marched at 6 a.m., heading through Winnsborough, and then they tore up railroad about three miles from town and destroyed about two miles of track working until dark and then moved back toward town two miles and bivouacked for the night.

February 22, 1865 The Third Brigade along with the Second Brigade continued to destroy the railroad track toward White Oak Station for about 7.5 miles and then rejoined the corps at Wateree Church. General Barnham, who commanded the Third Brigade, reported, "Marched at 7 a.m. toward White Oak and destroyed railroad until 4 p.m., when I received orders to move the Second Brigade with my own to Wateree Meeting House, which I did, arriving at 9 p.m. and reported to General J.W. Geary, commanding Division. The command marched fifteen miles and effectively destroyed three miles of railroad, every rail being twisted." WOTR Vol 47 p. 751

February 23, 1865 The Second Division marched at 6:30 a.m. and went through Morgan's camp and two miles since leaving Wateree Church took a road to their right for Rocky Mount Post Office on the Catawba River. General Geary reported, "This river crossing was one of the most difficult imaginable. The river was 250 yards wide and spanned by a single pontoon bridge. At the end of the bridge the steep, narrow road wound up a very high hill, which the trains after crossing ascended with great difficulty and only by the assistance of the troops. The soil everywhere was treacherous and the roads were deep and mirey." WOTR Vol 47 p. 688

The Second Division began to cross at 5:45 p.m. on a dark and rainy night and finished around 10 p.m. General Barnham reported, "Marched at 6.30 a.m.; crossed Wateree River at Rocky Mount. The approach to the river was very bad. The men were placed along the sides of the wagons to help them up a very steep hill. Every wagon received assistance from the men. Got into camp at 12 midnight. Day's march, twelve miles." WOTR Vol 47 p. 755

February 24, 1865 The Second Division only traveled four miles that day, as they had to corduroy the roads the whole way as the rain continued and other troops ahead of them slowed them down. It became very cold that night. The Third Brigade camped near Colonel Ballard's.

February 25, 1865 The Second Division stayed in camp that day, as it rained cold and heavy until midnight, and they were delayed by the 17th Corps who had taken the wrong road and were in front of them.

February 26, 1865 The Second Division, in the center of the corps, marched at 7 a.m. and followed the Third Division and went three miles to Russell Hill where they kept to their left and reached Hanging Rock Post Office and camped there. General Geary reported, "The weather today was warm and clear. Two-thirds of the road had to be corduroyed for our trains. In most place fence rails were abundant, and were quickly brought into requisition. The surface of the country since leaving Catawba River is a crust with quicksand underneath. Wagons and animals everywhere except on the corduroy broke through the crust to the depth of three feet or more. Hanging Rock Post Office is near a creek of the same name. Near the ford where the main road crosses is a large projecting rock on the hill-side overhanging the stream, and giving its designation. The place is noted as the scene of one of the minor conflicts of the Revolution, with which this State abounded in the days of Marion, Sumter, Cornwallis, and Tarleton; distance to-day, nine miles." WOTR Vol 47 p. 688

Research shows a battle occurred there on August 6, 1780, in which about 600 militia under Colonel Thomas Sumter destroyed the British camp and killed and wounded over 200 of the British troops under Major John Carden with a loss of 40 killed and a few wounded.

February 27, 1865 The Second Division crossed Hanging Rock Creek by a good ford with a smooth rocky bottom and then camped two miles beyond that at Ralling's farm. They only went three miles that day, as the soil was treacherous and full of quicksand.

February 28, 1865 The Second Division marched in front of the corps disencumbered toward Little Lynch's Creek and crossed over it on a good bridge and camped at noon near Clyburn's store. General Barnham reported, "Marched at 6.30 a.m. Road bad; most of it had to be corduroyed. The brigade, second in line, reached Little Lynch's Creek at 11 a.m., when I received an order from General J. W. Geary, commanding division, to move back and assist the wagon train through. I marched back nearly three miles and in rear of the First Brigade, which had been following my command. I assisted the wagons both by lifting them out of the mud and in building corduroy roads. Got into camp at 4 p.m. near Clyburn's store. Day's march nine miles." WOTR Vol. 47 p. 752

March 1, 1865 The second division was now in the rear of the corps and didn't move until 11:40 a.m. They crossed Big Buffalo Creek and then Lynch's Creek where they found a good bridge at Miller's Mill. Although it slightly rained that day the roads were generally good but had to be corduroyed in certain places. The Third Brigade marched about twelve miles that day and camped at Brewer's farm.

March 2, 1865 The Second Division was again in the rear

and didn't leave camp until 9:30 a.m. they marched due east on a road that intersected at Johnson's farm and the main Camden and Chesterfield road. They reached Big Black Creek but were halted for the night due to troops in front of them waiting for a bridge to be finished.

March 3 1865 Sherman entered North Carolina. Days later, Confederate forces under Bragg and Maj. Gen. Wade Hampton conducted small offensives at Wyse Fork and Monroe's Crossroads but with little effect on Sherman's campaign.

On **March 3, 1865**, Union General William Tecumseh Sherman and his Army of the Cumberland enter the Confederate State of North Carolina, having come from the recent destruction of Columbia and working their way north. Sgt. Moses Whitbeck is part of the 102nd NY Volunteers, which for the Carolinas Campaign has been part of 20th Army Corps, led by Brigadier General Alpheus Williams. Working their way north, Williams' corps is on the west heading toward Monroe's crossroads. On the east is Major General Howard's XI Corps. Sherman is trying to keep his destination secret by having the Corps split, but they are headed toward Fayetteville and a rendezvous with Confederate General Joseph E. Johnston and the battered Confederate Army of Tennessee. The Confederate army was organized into three corps, commanded by Lt. Gen. William J. Hardee, Lt. Gen. Alexander p. Stewart, and Lt. Gen. Stephen D. Lee. In between the two Union armies was Lt. Gen. William J. Hardee and his remnants of Confederate troops unsuccessfully trying delaying actions against Sherman's armies. Most of the Confederate regiments and division were now just mere fractions of what they had been through wartime devastation of those killed in battle, those who were captured, and those that died of wounds or disease. The South is almost defeated, and the Union men and Moses Whitbeck can feel it.

The Union Army 20th Corps marched at 6:30 a.m. and was delayed an hour repairing the bridge over Big Black Creek, which was in very bad condition. When they had reached the creek they found that the trains from the Third Division had not yet left so they crossed where Smith's Mill Creek was. The road from Smith's Mill Creek was in bad condition, and they had to corduroy a large part of it. They reached Chesterfield Courthouse at 9 p.m. and camped. The Third brigade was in the rear and got into camp at 11 p.m. having marched 14 miles.

March 4, 1865 The Second Division was in the center of the corps and marched at 7 a.m. They crossed Abram's Creek, Little Westfield and Big Westfield Creeks and camped near Sneedsborough, covering the plank road which runs from Wadesborough to Cheraw. The roads were in bad shape and had to be corduroyed, but they did manage to cover 10 miles that day. The men of the division foraged and found abundant supplies in the area between themselves and Wadesborough. Two grist mills were taken possession of and a large quantity of corn was ground for the use of the men.

March 5, 1865 The Second Division remained in camp as the pontoon bridge builders, protected by the 14th Corps were laying down a pontoon bridge at Pegue's Ferry on the Great Pedee River. The Third Brigade remained in camp while the 149th NY were engaged in corduroying the roads. Another mill was taken and run for the benefit of the command.

March 6, 1865 The Second Division marched at 8 a.m. taking the plank road for Cheraw and arriving at 12.30 p.m. They waited for the 15th Corps to cross the pontoon bridge there and then crossed themselves. They followed the

Fayetteville Road and then camped six miles from Cheraw at Smith's Mill on Wolf Creek, having traveled 15 miles that day.

March 7, 1865 The Second Division marched at 6 a.m. and led the corps. They went on good roads for a change, but the country was sandy and lots of resin was manufactured in the area. They marched toward Station 103 on the Wilmington, Charlotte and Rutherford railroad, destroying some of the track and some new iron rails that were stacked up and ready to ship to other places. General Geary reported, "Several large resin factories along our route were destroyed to-day. One alone contained 2,000 barrels of resin lately manufactured. A party of my foragers, combining with others from the Fourteenth Corps entered Rockingham to-day, driving out Hardee's rear guard, with whom they were actively skirmishing when Kilpatrick's advance reached the town. Weather clear and delightful; distance, thirteen miles." WOTR Vol.47 pt. 1 p.690

General Barnham from the Third Brigade that the 102nd NY Volunteers were part of, reported that at 11 a.m. the brigade crossed the boundary between South Carolina and North Carolina and camped at 1 p.m. after traveling 13 miles.

March 8, 1865 The Second Division was last in the Corps, and they didn't march until 11:45 a.m. They crossed the railroad and went on a small settlement road toward McFarland's Bridge on Lumber River. At 1 p.m. they reached Mark's Creek where they were halted until 4 p.m. by wagons of the Third Division. General Geary reported, "Finding the crossing here very swampy and almost impassable, I had it corduroyed, and passed my troops and trains across the creek before dark. As we proceeded the roads became very bad. A heavy rain had fallen all day, many swampy streams had to be crossed, and the soil elsewhere was full of quicksand. My labor was expended bringing my trains through. Passing the Fourteenth Corps in camp, I encamped three-quarters of a mile in rear of the Third Division; distance, seven miles." WOTR Vol. 47 pt. 1 p. 690

March 9, 1865 The Second Division marched at 6:30 a.m. and after crossing a small swamp were stopped again by the wagons of the Third Division. General Geary had his men make a road and a bridge to the left of the main crossing and passed over without difficulty. They then moved forward over mostly solid ground and open woods. When they got within three miles of the Lumber River, they came to another stream where the trains from the First and Third Brigades were in their way. General Geary found a road that passed to the right and crossed the creek by its head. They took it and continued until they reached McFarland's Bridge where they stopped and camped. A cold rain had set in about 3 p.m., and they had to corduroy the last two miles. Total distance traveled was 12 miles. The Third Brigade under General Barnham was assigned to guard a large pontoon train from the Fourteenth Corps and help them through the Lumber River crossing. They then marched forward corduroying the road until they reached Crossed Hill's Creek and camped there having traveled 8 miles.

March 10, 1865 The Second Division had another tough day of travel with bad roads and wagon trains in front of them slowing them down. They crossed Buffalo Creek and a couple of small swamps and about a mile past they were stopped again by wagons of the Third Division, so they camped there for the night having only traveled 3.5 miles.

March 11, 1865 Union General William T. Sherman's forces entered Fayetteville, North Carolina, facing little resistance, and took command of the city. The destruction of the bridge over the Cape Fear River angered Sherman and delayed his advance. While preparations were made to cross the river, Sherman sent the wounded soldiers and all the Southern refugees to Wilmington. Fayetteville was treated harshly for the destruction of the bridges, the armed resistance when Union soldiers first arrived, and because the city was the location of a Federal arsenal before the war. Sherman rested his army for one day and then resumed his trek toward Goldsboro. Upon leaving the city, Sherman ordered the destruction of specific structures within Fayetteville. However, much more was destroyed than initially ordered. North Carolina History Project: Carolinas Campaign (January 1865-April 1865)

While Sherman and his men faced little military resistance in Fayetteville, their goals and actions remained the same – to destroy Confederate resources and its will to fight. Just prior to entering Fayetteville on March 10, General Frances Preston Blair released Special Order No. 63 outlining the foraging practices for his men in the city of Fayetteville. In the order, Blair reiterated that "The State of North Carolina is to a great extent loyal," but they had to forage what was necessary in Fayetteville. The men who did the foraging, Blair described, often went by the name of a "bummer." Lieutenant Colonel George Nichols of the Union Army wrote fondly of the bummers while he was in Fayetteville. According to Nichols, a bummer "is a raider on his own account—a man who temporarily deserts his place . . . and starts out upon an independent foraging expedition." However, Nichols acknowledged, "these wanderers from the ranks are often a great benefit to the army." Although bummers took mass amounts of food from the Fayetteville population, Sherman and his men believed that bummers played a vital role in keeping the Union Army fed and advancing. Bummers derived their importance from Sherman's destructive strategy. He had largely destroyed much of the transportation infrastructure on his march through the South in order to cut off supply lines, making foraging from local populations a military necessity. Excerpts taken from Civil War Era NC: Fayetteville, Union Accounts

Sherman's army moved beyond the taking of food to desolate the town of Fayetteville. The Union troops destroyed many key buildings and factories within the town. Sherman ordered his men "to destroy all railroad property, all shops, factories, tanneries, and all mills, save one water-mill of sufficient capacity." The main item that Sherman wanted to destroy was the arsenal building, which the United States federal government had built in Fayetteville before North Carolina seceded. Sherman said of the arsenal in Fayetteville, "The enemy shall not have its use, and the United States should never again confide such valuable property to a people who have betrayed a trust." Letter from William Sherman to Ulysses Grant, March 12, 1865

The destruction of the arsenal held importance as a military item, but it was also important symbolically as Lieutenant Colonel George Nichols wrote in his diary, "We shall destroy it utterly . . . . By Monday night that which should have been the pride and honor of the state and the country will be a shapeless mass of ruins." The de-struction Sherman brought to Fayetteville intended to cripple not just resource centers like factories but also to cripple its pride.

As for the Union Army 20th Corps Second Division on **March 11, 1865**, they marched at 6.30 a.m., crossed Nicholson's Creek, and were stopped about a mile west of Rockfish Creek when they reached the First Division's camp. Here the Second Division was tasked with leading the entire group of trains for the corps toward Fayetteville as they were heading for Fayetteville disencumbered. This was about 1,000 trains spread out among the three brigades of Second Division. They crossed Rockfish Creek, Beaver Creek, Puppy Creek and a number of smaller streams. Some of the roads were bad, especially near the stream crossings and the men had to corduroy those places. At Lamont's Mill on Puppy Creek, they took a small road to their left which after a mile ran into the Fayetteville and Albemarle Plank roads. Here they camped for the night, still 13 miles from Fayetteville.

General Barnham of the Third Brigade reported, "marched at 6 a.m.; overtook the First Division of this corps at 11 a.m. at Rockfish Creek. The Brigade was then assigned to take charge of 500 wagons; crossed Towny, Beaver Dam, Rockfish, Cat Tail, Beaver Black Branch, and Puppy Creeks; bivouacked at 8 p.m. Day's march 15 miles." WOTR Vol 47 p. 752

March 12, 1865 The Second Division was on the march early at 5 a.m., taking the plank road to Fayetteville and reaching it at 1 p.m. They camped southwest of the town for the night. General Geary reported, "The road to-day, which crossed several creeks, was generally good. At Fayetteville we found communication with Wilmington temporarily opened by gun boats which had ascended the Cape Fear river. Sent out a mail for the North this afternoon." WOTR Vol. 47 pt. 1 p. 691

March 13, 1865 The Second Division marched at 2 p.m. and passed through Fayetteville in what is called Order of Review in which Major General Sherman reviewed them from the balcony of the principal hotel in town. They keep marching and crossed the Cape Fear River on a pontoon bridge and camped three miles beyond on the low flat ground along the Smithville Plank Road having traveled five miles.

March 14, 1865 The Second Division stayed in camp. This

was a welcome reprieve for most of the men. However, an officer of the 29th Pennsylvania, Lieutenant Ethan O. Fulce, was killed during a foraging mission.

On **March 15, 1865**, the Battle of Averasboro (also called Averysborough, Smith's Mill and Black River) took place near Fayetteville. That afternoon, Major General Judson Kilpatrick's Union cavalry advanced up the road and skirmished with Confederate General Hardee's lead elements before withdrawing and requesting infantry support. Just after midnight, Union troops began concentrating south of Hardee's first line.

As for the Second Division, they were tasked with guarding and escorting the wagons for the entire corps and missed the action as they were in the rear. They were busy corduroying the bad roads so they could get the wagons through, but they did manage to cover 10 miles. The Third Brigade marched at 8 a.m. and were in front of the Division and therefore unencumbered and repaired a large portion of the road as they went. When they were three miles from Black River they stopped for the day but sent a detachment of 100 men of the 111th Pennsylvania to the river to occupy and hold the crossing. At 6 p.m., General Geary ordered General Barnham to send four regiments to the river and they arrived there at 8 p.m. They camped there near the river after crossing Horse Pen Creek and covering 12 miles.

At approximately 9 a.m. on **March 16, 1865**, soldiers from the Union 14th and 20th Corps began advancing up the road toward Rhett's Brigade. Greatly outnumbered, Rhett's men could not stem the Union assault, and they fell back to Elliot's line. About 1 p.m., the Union moved against Hardee's second line. Once more they overwhelmed the confederates, forcing them to fall back. Bolstered by their success, the Union troops moved forward against the last rebel line. Unlike the green troops who occupied the first two lines, McLaws' battle-hardened confederates held their position. A steady rain that had fallen throughout the day turned into a downpour. Ohio Civil War Central.com: Battle of Averasboro

Union Generals Sherman and Slocum were preparing for a final assault on Hardee's third line, but the road and surrounding areas turned to mud, hindering troop movements. With nightfall approaching, Sherman postponed the attack until the next morning.

The Second Division had missed the action at Averas-borough as they were guarding the wagon trains. Part of the Third Brigade crossed Black River and were sent forward to protect the crossing, the rest of the brigade was busy cord-uroying roads. Orders came from General Geary to send the 137th and 149th NY Regiments to join General Pardee commanding First Brigade to help protect the wagons. The bridge was completed and the command crossed over around 11 a.m. At 1 p.m. the brigade was threatened by rebel calvary on their left flank so the 60th NY Regiment was put in position to defend. No attack was made and the brigade moved on and camped near the 15th Corps having traveled four miles.

General Sherman anticipated a major assault against the Confederate third defensive line at dawn on **March 17**. The defending Confederate General William Hardee, having succeeded in his mission of delaying and disrupting the advance of General Sherman's left wing, withdrew his troops under cover of darkness the night of **March 16** and conducted a forced march to the vicinity of Bentonville. There these Confederate veterans of the Battle of Averasboro would join

General Johnston and his army to surprise the Union 14th and 20th corps and begin the Battle of Bentonville on **March 19, 1865**.

As for the Second Division of the 20th Corps, because they were blocked in moving by the 15th Corps, they stayed in camp but sent wagons forward with bread and coffee for the wounded. The 102nd NY Veteran volunteers were sent out with a small wagon train to forage for supplies but was not very successful. WOTR Vol 47 p. 753

March 18, 1865 The Second Division marched at 6 a.m., taking the Tarborough Road which was a direct route to Bentonville. The road was in bad condition, and the Division crossed several streams and camped 1.5 miles east of Rainers' Mill on Seven-Mile Creek. They spent the whole day corduroying all of the road they travelled that day having covered 8.5 miles. The Third Brigade marched at noon and crossed the Little Cohera Creek and reached camp between 10 p.m. and 4 a.m. the next morning.

March 19, 1865 Battle of Bentonville Carolinas Campaign

Confederate General Johnston saw an opportunity to strike at Union General Sherman's forces before they reached Goldsboro. He planned a major effort at the crossroads town of Bentonville, taking advantage of a gap between Sherman's widely separated forces. His hope was to destroy one wing before the other could arrive and support it. Fighting raged at the Battle of Bentonville from **March 19-21**, and the Confederates came close to success, but Union troops held on with dogged determination.

On **March 19, 1865**, In the aftermath of the Battle of Averasboro, Union General William T. Sherman continued his march through the Carolinas, destroying railroads and disrupting supply lines on its way to join Lt. Gen. Ulysses S. Grant's army near Petersburg and Richmond. On this day, as the respective Federal wings approached Goldsboro, Maj. Gen. Henry W. Slocum's wing encountered Johnston's hodge-podge army. Confederate General Johnston saw an opportunity to strike at Union General Sherman's forces before they reached Goldsboro. Johnston's forces concentrated at Bentonville with the hope of falling upon Slocum's wing before the Federal wing of Maj. Gen. Oliver O. Howard could come to Slocum's support. Union troops of the XIV Corps of the Army of Georgia (previously called the Army of the Cumberland) encountered Confederate forces at Bentonville, as they marched toward Goldsboro. This Confederate resistance stiffened as Union Army commander Major General Henry Warner Slocum had no idea that Johnston's entire Confederate force was in front of him. The fighting was fierce, and the Union lines were driven back. Eventually the 20th Corps, including the 102nd NY Volunteers and Sgt. Moses Whitbeck, arrived to launch an attack to clear the way.

It is interesting to note that afternoon the Confederate Army of Tennessee launched its last attack of the war, crushing Union resistance and at one point surrounding an entire division. One Confederate noted that the attacking force "looked like a picture and was truly beautiful." Yet, he admitted it was "painful to see how close their battle flags were together, regiments being scarcely larger than companies." Emerging Civil War.com. The Last Charge of the Army of Tennessee, original quote can be found: Mark Bradley, Bentonville (Campbell, CA: Savas Beatie, 1996), p. 204.

As the attack overwhelmed the Union troops of Brigadier General William Passmore Carlin's Division, one soldier noting that they "stood as long as man can stand." : Mark Bradley, Bentonville (Campbell, CA: Savas Beatie, 1996), p. 212

Nearby, Brigadier General James Dada Morgan's Division fought from both sides of their earthworks, as they beat back Confederate assaults from front and rear. By late afternoon, the Confederates pushed the surprised Union forces back. The Confederates reached their high water mark at the Morris Farm, where Union forces formed a defensive line. After several Confederate attacks failed to dislodge the Union defenders, the rebels pulled back to their original lines. Nightfall brought the first day's fighting to a close in a tactical draw. Pillar to Post: RETRO FILES / CIVIL WAR AT THE BITTER END.

As for the 20th Corps Second Division, they marched at 6 a.m., still following the Tarborough Road to Newton Grove Post Office at Dr. Monk's house. Then they turned right on the Goldsborough new road and camped at Canaan Church. The 15th Corps was within a mile from there and delayed the 20th Corps from moving until 4 p.m. When they did move the road was bad and they corduroyed it for three miles. About 10 p.m., General Geary received orders to send all their remaining ambulances as well as all the empty wagons of the corps and the ammunition and supply wagons. At midnight, General Geary received orders to take two brigades and Sloan's artillery battery and join the rest of the corps by daylight as the enemy had been reinforced and a heavy attack was expected in the morning. The First and Third Divisions had been heavily engaged and the Second Division was called for support as they were guarding wagon trains.

The next day, **March 20, 1865**, both sides held their ground, as the Army of the Tennessee arrived to bolster Slocum's forces. Union General Howard's right wing arrived to reinforce Slocum, which put the Confederates at a numerical disadvantage. Sherman expected Johnston to retreat and was inclined to let him do so. Although Johnston began evacuating his wounded, he refused to give up his tenuous position, guarding his only route of escape across Mill Creek. Outnumbered, his only hope for success was to entice Sherman into attacking his entrenched position, something Sherman was unlikely to do. Heritage Post.org: The Battle of Bentonville

As for the Second Division, General Geary was on the move with the force ordered and reached the main part of the corps at 4.30 a.m. They had traveled a swampy road for eight miles. The Division was put in reserve behind the left part of the Union line. The expected attack did not occur and the troops stayed where they were quietly resting. At 11 a.m., orders came from General Williams commanding the Corps to move the trains toward Goldsborough, crossing Falling Creek near Irvine King's house. General Barnham of the Third Brigade reported that the place where they were resting earlier in the day was called Harper's Farm.

On **March 21, 1865**, fighting resumed in the last major Civil War battle, which was fought at Bentonville, North Carolina, with a limited Union attack on the Confederate right. Confederate General Johnston remained in position, and heavy fighting erupted south of the Goldsboro Road in an area later called the "Bull Pen" between Morgan's and Hoke's men. Under a heavy rainfall, Union Maj. Gen. Joseph A. Mower led a "little reconnaissance" toward the Mill Creek Bridge. When Mower discovered the weakness of the Confederate left flank, Mower launched an attack against the small force holding the bridge. A Confederate counterattack, combined with Sherman's order for Mower to withdraw, ended the advance, allowing Johnston's army to retain control of their only means of supply and retreat. Excerpts taken from Heritage Post.org: The Battle of

Bentonville

As for the Second Division, they remained where they were massed in reserve, as the rebels had left that area. Colonel Mindil, from Second Brigade, was ordered to move the wagon trains near the railroad crossing of the Neuse River and for a temporary depot to be established. From there he was directed to send wagons to Kinston for supplies. The 149th NY was sent in the direction of Cox's bridge on the Neuse River to be an escort to the trains that were carrying the wounded of the corps. At 6 p.m., the 111th Pennsylvania Volunteers from Third Brigade were sent with the pioneers of the brigade in the same direction to repair the road. Heavy rain set in that evening and lasted until midnight.

Bentonville was the largest battle ever fought in North Carolina, involving more than 80,000 troops (60,000 Union and 20,000 Confederate). During the battle, the Confederates suffered a total of nearly 2,600 casualties, 239 killed, 1,694 wounded and 673 missing. About half of the casualties were lost in the Army of Tennessee. The Union Army lost 194 killed, 1,112 wounded, and 221 missing, for a total of 1,527 casualties. The wounded were treated at the house of John Harper, with 360 unknown Confederates buried in a mass grave next to the Harper family cemetery. Bradley, Mark L. Last Stand in the Carolinas: The Battle of Bentonville. p. 403-404

During the night of March 21 until the following dawn, Johnston withdrew his army across Mill Creek and burned the bridge behind him, leaving behind a cavalry detachment as a rearguard. The Union Army failed to detect the Confederate retreat until it was over. Sherman did not pursue the Confederates but continued his march to Goldsboro. Bradley p 400-401

A more aggressive Union attack at Bentonville might have

proved decisive, as the Confederates had only one line of retreat over Mill Creek. Yet Sherman was more focused on uniting the various Union columns at Goldsboro, and let the Confederates slip away.

Pulling back, both armies rested and prepared for the next step. Near Smithfield, Johnston reorganized the army on April 9, the same day that General Lee met with General Grant in the McLean house at Appomattox Courthouse. The army had seen other consolidations, but this was more drastic than anything experienced before. The army had too many undersized regiments and an overabundance of general officers. Many regiments were combined into new Consolidated Regiments. Johnston ordered that the massive reorganization move "with all possible speed, Sherman will not give us much rest." In another order, he wrote that consolidation begin "without delay."

After North Carolina's largest battle, the Confederates were defeated in the last major battle of the war.

Goldsborough was the scene of another Union offensive in 1865, during Union General Sherman's Carolinas Campaign. After the battles of Bentonville and Wyse Fork, Sherman's forces met with the armies of Schofield, their troops taking over the city in March. During the following three weeks, Goldsborough was occupied by more than 100,000 Union soldiers. After the war was over, some of these troops continued to stay in the city. Wikipedia: Goldsboro, North Carolina

March 22, 1865 The 102nd NY Veteran Volunteers, under Lieutenant Colonel Chatfield, as part of the Third Brigade of Second Division, was sent in charge of a train, comprising of pack-mules, headquarters trains, a few ammunition wagons and ambulances, with orders to cross Falling Creek by the same rout taken by the other trains. Shortly after that it was discovered that the enemy had evacuated during the night. At 8 a.m., General Geary and the Second Division moved ahead of the corps and used his troops to repair the roads and get the wagons through. They camped on the east side of Falling Creek and the wagons on the west side, having made 15 miles that day.

With Johnston out of the way, Sherman reached Goldsboro on **March 23, 1865**. The addition of the Army of Ohio swelled the size of Sherman's forces to nearly 90,000 soldiers.

The Second Division marched in the front of the corps and went by a direct road to Cox's Bridge, passing Falling Creek Post Office and Grantham's Store. At 10 a.m., they reached Cox's Bridge, but the 14th Corps was ahead of them, so they had to wait their turn. At noon, they began crossing. Taking the right-hand road after crossing the Neuse River toward Goldsborough, they camped three miles from Cox's Bridge at the junction of the Smithfield Road. They skirmished with a considerable enemy group of calvary holding on their left flank but received no casualties. Then they camped having traveled 12 miles. The 102nd and 149th NY Regiments rejoined the Third Brigade and camped near Beaver Dam Creek.

March 24, 1865 The Second Division was ordered to send all their wagons at 2 a.m. toward Goldsborough so they could get through by daylight. The Division marched out in the rear of the corps at 7 a.m. and crossed the Raleigh railroad and two large creeks and at noon passed in order of review through Goldsborough in front of General Sherman. And then went one mile north and camped near the Weldon railroad.

All the wagons and animals of the Third Brigade were sent to Goldsborough at 2 a.m. as well, and the brigade moved out at 7:30 a.m. They crossed Little River and passed through Goldsborough at 11 a.m. in column by companies and were reviewed by Major General Sherman, then they moved outside of town and camped on the north side having traveled eight miles.

March 25, 1865 The Second Division was moved to a more permanent position to camp on the right side of the corps. Here they were directed to make comfortable camps and measures were taken to procure for the troops full supplies of food, clothing and for the men to get themselves cleaned up and returned to proper appearance after a lengthy campaign.

Lieutenant Chatfield who commanded the 102nd NY Veteran Volunteers wrote of the Carolinas Campaign, "During the whole of this campaign, although participating in no engagement, the men have always evinced a readiness to undertake any labor or danger, and have undergone hardships and fatigue second to none heretofore required of them. From first to last, and with but two exceptions, not a day has passed but that this regiment has been with the wagon train and assisting it over roads. Constant and weary labor has been demanded and given in corduroying and repairing the roads, which from the beginning have been of the worst character and rendered nearly impassable by the almost constant rains which have been prevalent during the campaign. The only railroad destroyed by the regiment was that near Winnsborough, on the Columbia and Charlestown railroad, where a little more than a mile was most effectually destroyed, every rail having been twisted... Throughout the lengthy and laborious campaign both officers and men have behaved well and at all times cheerfully complied with whatever has been required of them. They have arrived at this point deficient of clothing of every kind and in great need of a rest from their heavy labors." WOTR Vol 47 p. 761-762

From **March 25th until April 10**, the 20th Corps stayed in camp near Goldsborough. They were refitting and getting cleaned up from the Carolinas Campaign.

On **March 27, 1865**, General Grant summoned General Sherman to his headquarters at City Point, Virginia, to meet with President Lincoln. Expecting the imminent fall of the Confederacy, the three men discussed procedures and terms of surrender for the rebel armies remaining in the field. AmericanHistoryCentral.com: Carolinas Campaign

April 5, 1865, The Union Army reorganized the 20th Corps with the 14th Corps into a new department, The Army of Georgia. Union General Alpheus Williams was returned to the First Division, as Major General Joseph "Fighting Joe "Mower was put in command of the new Army. General Williams considered resigning from the Army, but there was much pressure from his old division to come back, so he did. He also wrote that Union General Sherman would be mad at him if he resigned. General Williams wrote to his daughters, "I was gratified yesterday, when three Brigade Commanders called on me formally and announced that they spoke the wishes of every officer and man in the division that General Williams should resume command of their Division." FTCM p. 380

General Williams wrote that Major General Mower was a very pleasant, gentlemanly man of the old army and had been with Union General Sherman during all his Mississippi campaigns. He also wrote that he preferred his old division to any corps except the 20th. As for the 102nd NY Veteran Volunteers, they were still in the Third Brigade under General Barnham and the Second Division under General Geary.

April 6, 1865 The 102nd NY Volunteers received news of the evacuation of Richmond, forced by Union General Grant's Army. Union General Alpheus Williams wrote, "The regiments are cheering all around me. I do not feel so much rejoiced. I think if Lee had held on a little longer it would be better for us, as we should have made a junction with Grant. Now the whole rebel army I fear, will get between us." FTCM p. 380

Although Richmond had been captured, Confederate Generals Robert E. Lee and Joseph Johnston still led armies in the field. General Geary would write, "Since we have been in North Carolina I have seen some very terrible things. The people here are nearly all in a starving condition, food is so scarce that many people are starving." APGTW p. 236

General Alpheus Williams would write that the only people left in the South were "women and children, old men and the decrepit, and deserters from the Confederate Army of which there were many." FTCM p. 381

April 8, 1865 The news of the success at Richmond against Confederate General Robert E. Lee has been confirmed. General Geary wrote, "We have the full news of the glorious success at Richmond against Lee. Every body in the army of course is rejoicing, and at this moment every thing is in motion to march early tomorrow morning in the direction of Raleigh, with a view to intercepting the rebels in their westward flight. God grant that this war may soon be over. Every thing looks well here, and the blessings of heaven seem to rest upon our cause... Thank God, who has given us, who have stood the brunt of this war, such cause of rejoicing and hopes of speedy and glorious success." APGTW p. 237

April 10, 1865. The Union Army of Georgia left Golds-

borough led by General Alpheus Williams' division. They passed through the town and recrossed Little River, then passed
over the same road they went into Goldsborough on, then they
went as far as Beaver Creek, then north on the Smithfield
Road. They crossed Moccasin Creek after much opposition by
Confederate calvary.

General Alpheus Williams wrote, "Indeed, we skirmished
all day, but drove the enemy rapidly. At the Moccasin, which is
a broad swamp with two deep streams, we were obliged to
travel the men to their arm pits to get possession of the bridges
which had been thrown into the creek." FTCM p. 381

The division lost two killed and three wounded. General
Williams himself was in danger near one of the bridges, looking
after the skirmish line.

As for the Second Division commanded by General Geary,
General Pardee who commanded First Brigade was ill from the
Carolinas Campaign, so Colonel Mindil from the 33rd New
Jersey temporarily took his place. Brigadier General P.H. Jones
rejoined the Division at Goldsborough and took command of
the Second Brigade. The Third Brigade that included the
102nd NY Veteran Volunteers was still led by General
Barnham. The reorganization cost the Division possession of
36 wagons that went to the 23rd Corps. The Second Division
was second in line of march behind the First Division. They
marched a total of fourteen miles that day under drizzling and
very disagreeable conditions. They camped at Thomas
Atkinson's plantation, one mile north of Moccasin Creek.

Major O. J. Spaulding of the 102nd NY Veteran Volunteers,
who took over for Colonel Chatfield from April 10 until May
11th, reported, "Pursuant to orders received from headquarters Third Brigade, Second Division, Twentieth Army

Corps, this regiment left camp at 5 a.m., April 10, 1865; march-ed through Goldsborough on the Smithfield road, a distance of nine miles, and bivouacked for the night." WOTR Vol 47 p. 763

April 11, 1865 Union General William Tecumseh Sherman returned from a meeting with Grant and Lincoln to Goldsborough, and planned a move against the North Carolina state capital of Raleigh. The Union Army of Georgia with the 102nd NY Veteran Volunteers and Sgt. Moses Whitbeck marched from Goldsboro to Smithfield without much opposition. The Second Division was in the lead, and when they got to the large, covered bridge over the Neuse River it was on fire, and they had to install pontoons to get across.

General Geary reported, "Marched at 5.30 a.m., my Second Brigade leading. Skirmished nearly all day with the enemy's cavalry, driving them rapidly. No casualties on our side. Encamped at 2 p.m. at Smithfield, from which place the enemy's rear guard retired this morning, burning the bridges across Neuse River. We crossed during the day Boorden's and Pole Cat Creeks. At Boorden's Creek the bridge, sixty-five feet long, was destroyed. With two companies of my pioneers it was rebuilt in seventeen minutes; weather to-day mild, with occasional showers. Country well cultivated and containing a number of handsome residences for this region; distance, eleven miles and a half." WOTR Vol. 47 pt. 1 p. 700

General Alpheus Williams wrote, "The country between Goldsboro and Smithfield is much better that any we have seen in North Carolina, better soil and cultivation. The houses are of better character and the people at home generally profess to be Union and rejoice at the probable end of the war." FTCM p. 381

April 12, 1865 The Union Army of Georgia men received word that General Robert E. Lee had surrendered the Army of Northern Virginia to Union General Grant at Appomattox Courthouse, Virginia. Union General Alpheus Williams and his 20th Corps was in the rear and before they crossed the Neuse River, they heard of the dispatch to General Sherman telling of the news. Williams wrote, "I have never seen Sherman so elated. He called out to me from a bevy of mules and as soon as I could reach him through the kicking animals he grabbed my hand and almost shook my arm off, exclaiming 'Isn't it glorious? Johnston must come down now or break up!' I confess that I felt and expressed a pretty large sized 'Laus Deo' at the prospect of an early end of this great rebellion and a return to my family. Our long and tedious marches, for now nearly a year, with but little cessation, has quite filled me with a yearning for quietude and repose." FTCM p. 381

The 102nd NY Volunteers as part of the 20th Corps Second Division marched this day on the left road to Raleigh, twice crossing Swift Creek. It was a humid, showery day, and the men were tired and inclined to straggle along. They camped that night at Mrs. Saunder's plantation, having traveled 14 miles.

Sherman moved on and occupied Raleigh on **April 13, 1865**.

At daylight on **April 13, 1865**, the 102nd NY Volunteers and the Union Army of Georgia 20th Corps set out for Raleigh. Because the 14th Corps was ahead of them and in their way, they took crossroads and byroads toward Raleigh. They reached the southwest side of Raleigh after a 12-mile march and found out the rebels were gone and Union General Hugh Kilpatrick's Calvary was in possession of Raleigh. They continued their march another four miles and camped near the

Raleigh Asylum for the Insane. Union General Alpheus Williams wrote, "We passed through some fine country today, our march over sixteen miles. We encamped near the Insane Asylum. While waiting orders General Mower and myself visited the inmates. Dr. Fisher, the superintendent, was greatly distressed at the presence of our troops, but as we saw a long line of rebels works running directly through the asylum grounds we concluded he had other motives. Our men gathered pretty close about the building and were greatly entertained by the Union speeches and songs of several of the inmates. We saw a good many strange cases inside, several of whom made quite eloquent talks of the old flag. Indeed, the prevailing sentiment of these insane seemed to be for the old Union." FTCM p. 382

Major Spaulding of the 102nd reported, "left camp at 5.30 a.m.; marched with but few halts during the day. Marched on Raleigh railroad about six miles from the city; passed to the left of the town and bivouacked at about 4 p.m. near lunatic asylum; marched a distance of about seventeen miles." WOTR Vol. 47 pt. 1 p. 763

Raleigh was a large, sprawling, well-built town with good water and a respectable State House in the center and many fine homes around it, but the Union Army found no signs of commerce or manufacturing.

When Confederate General Joseph E. Johnston met with Jefferson Davis in Greensboro on **April 12–13**, he told the Confederate president, "Our people are tired of the war, feel themselves whipped, and will not fight. Our country is overrun, its military resources greatly diminished, while the enemy's military power and resources were never greater and may be increased to any extent desired. ... My small force is melting away like snow before the sun." <small>Wikipedia: Carolinas Campaign</small>

April 14, 1865 A single shot rang out at Ford's Theater in Washington DC. The crowd heard the words "Sic semper tyrannis! (Ever thus to tyrants!) The South is avenged," as John Wilkes Booth jumped from the Presidential booth onto the stage he broke his leg but hobbled away and fled on horseback. President Abraham Lincoln had been shot in the head. The wound was mortal.

Learning that President Lincoln was to attend Laura Keene's acclaimed performance "Our American Cousin" at Ford's Theater on April 14, Booth plotted the simultaneous assassination of Lincoln, Vice President Andrew Johnson, Secretary of State William H. Seward, and Union General Ulysses S Grant. By murdering the president and two of his possible successors, Booth hoped to throw the U.S. government into disarray.

On the evening of April 14, conspirator Lewis T. Powell burst into Secretary of State Seward's home, seriously wounding him and wounding eight others.

Meanwhile, just after 10 p.m., Booth entered Lincoln's private theater box unnoticed, and shot the president with a single bullet in the back of his head from a .44 single shot pistol. Although Booth had broken his left leg jumping from Lincoln's box, he succeeded in escaping both the theater and Washington D.C. The president, mortally wounded, was carried to a cheap lodging house opposite Ford's Theater. About 7:22 a.m. the next morning, he died. He was the first U.S. president to be assassinated.

Booth was a well-regarded actor who was particularly admired in the South before the Civil War. During the war, he stayed in the North and became increasingly bitter when audiences weren't as enamored of him as they were in Dixie. Along with friends Samuel Arnold, Michael O'Laughlin and John Surratt, Booth conspired to kidnap Lincoln and deliver him to the South.

On March 17, along with George Atzerodt, David Herold and Lewis Powell, the group met in a Washington bar to plot the abduction of the president three days later. However, when the president changed his plans and did not show, the scheme was put on hold. Shortly afterward, the South surrendered to the Union and the conspirators altered their plan. They decided to kill Lincoln, Vice President Andrew Johnson, General Grant, and Secretary of State William Seward on the same evening.

When April 14 came around, George A. Atzerodt backed out of his part to kill Johnson and instead got drunk. Booth went to drink at a saloon near Ford's Theatre to kill some time until the appointed hour. At about 10 p.m. he walked into the theater and up to the president's box. Lincoln's guard, John Parker, was not there because many believe he had gotten bored with the play and left his post to get a beer. Booth easily slipped in to the Presidents' box and shot the president in the back of the head. The president's friend, Major Rathbone, attempted to grab Booth but was slashed by Booth's knife. Booth injured his leg badly when he jumped to the stage to escape, but he managed to hobble outside to his horse.

General Grant, just days after accepting Lee's surrender, accepted Lincoln's invitation to attend "Our American Cousin" at Ford's Theatre on the evening of **April 14**. The general's wife, however, had recently had a disagreement with Mary Todd Lincoln and didn't want to attend. Grant backed out, citing the couple's desire to travel to New Jersey to see their children. Lincoln had a difficult time finding a someone else to attend. Secretary of War Edwin Stanton, Speaker of the House Schuyler Colfax, and even son Robert Todd Lincoln turned down the tickets before Clara Harris, daughter of New York Senator Ira Harris, and her fiancé, Major Henry Rathbone, accepted. Rathbone attempted to apprehend Booth after he shot the President, but was cut with Booth's knife on his left arm from his elbow to his shoulder. Rathbone would survive, but the incident left him scarred for life mentally and he eventually killed his wife and was committed to an insane asylum in Germany.

As for the Union Army 20th Corps Second Division, they were unaware of what had happened with the President and were now camped near Raleigh and would stay there until April 24.

April 15, 1865 In the hours after Lincoln died in the back bedroom of William Petersen's boarding house, souvenir hunters ransacked the property and snatched numerous relics of the martyred president. Deciding to cash in himself, Petersen began to charge admission to the hundreds of curiosity seekers who came each day to see Lincoln's bloody deathbed, which incredibly continued to be slept in by tenant William Clark each night. Petersen fell into financial difficulty in 1871 and died after being found on the lawn of the Smithsonian Institution following an opium overdose. Christopher Klein: 10-things-you-may-not-know-about-the-lincoln-assassination.

With Confederate General Robert E. Lee's army defeated and the Confederate government in exile, Confederate General Joseph Johnston realized the hopelessness of his situation. Isolated and outnumbered three-to-one, Johnston contacted Sherman on **April 16** to discuss surrender terms. The generals met the next day at a farm known as Bennett Place, near Durham, N.C., where Johnston surrendered the 89,270 troops under his command in the Carolinas, Georgia, and Florida.

As for the Second Division, they were camping near Raleigh and waiting the results of negotiations between Generals Sherman and Johnston.

On **April 17, 1865**, the two generals met at James Bennett's 350-acre farms in Durham's Station, a rail stop between Raleigh and Greensboro. What evolved over the next nine days would prove to be one of the most extraordinary periods in our nation's history and also cemented an improbable friendship between the two generals, who had never met before in person. Historynet.com: All or Nothin': The surrender Sherman and Johnston crafted at Bennett Place.

As for the men of the 20th Corps they would receive official news today of the assassination of President Lincoln. General Geary reported, "The news produced the deepest grief and indignation throughout the command." WOTR Vol 47 pt. 1 p 700

Then General Geary wrote his wife, "We have just learned the assassination of the President, his untimely loss has created a profound sensation in the entire army, and if we have to fight anymore, woe to rebeldom! The cowardly assassins are only exhibiting the same phases which so greatly embittered me toward them ever since the Kansas affair, and no one understands better than I their nefarious designs." APGTW p. 239

General Williams wrote his daughter, "The news of the President's death came directly upon the news of Lee's surrender. I have never seen so many joyful countenances so soon turned to sadness." FTCM p. 383

On April 18, 1865 Near Durham, North Carolina, Confederate General Johnston and Union General Sherman met at noon at the Bennett House. Johnston, through Breckenridge, obtained Davis's authorization to surrender the remaining Confederate armies but wanted Sherman's explicit assurance for the protection of his soldier's Constitutional rights. Sherman assured him that Lincoln's 1863 Amnesty Proclamation and the terms of the Appomattox surrender allowed for a full pardon of all Confederate soldiers, from privates to the commanding general. Sherman and Johnston eventually reached an agreement. Under this agreement, hostilities would be suspended pending approval of the agreement. Confederate arms were to be deposited in the respective state arsenals and could only be used within that state, and officers and men had to sign an agreement to cease all hostilities of war. In addition, the President of the United States would recognize all Southern state governments as long as their officers and legislators took an oath of allegiance. The federal court system also would be re-established in the Southern states. The President would also guarantee the personal, political and property rights of the Southern people and grant legal amnesty to all Southerners, which implicitly included Davis and his cabinet. These terms were incredibly lenient to Southerners and followed Sherman's policy of a hard war followed by soft terms. Sherman did not want to punish the South but rather welcome them back into the fold of the United States with open arms to alleviate any resistance. Regardless his motivations, by offering these terms Sherman delved into political matters that he had no authority over.

Excerpts taken from: American Battlefield Trust: Bennett Place Surrender

As the sun set on **April 18**, a surrender document Sherman and Johnston had signed promised the end of four years of civil war. The agreement that the two men signed was the largest surrender of Confederate soldiers during the American Civil War.

April 22, 1865 The Union Army 20 Corps marched through Raleigh and were reviewed by General Sherman.

April 25, 1865 The Third Brigade broke camp at 8 a.m. and marched 12 miles on the Holly Springs Road to Jones' Cross-roads and camped there.

April 26, 1865 Johnston formally surrendered his army to Sherman at Bennett Place just outside of Raleigh.

Radical Republicans in Washington, embittered by President Lincoln's assassination on **April 14**, rejected the accord because it went beyond strictly military issues. Sherman and Johnston met again at Bennett Place this day and signed a new surrender document, using the same terms Grant and Lee had agreed to at Appomattox Courthouse. The signing of the new agreement brought Sherman's Carolinas Campaign to an end. AmericanHistoryCentral.com: Carolinas Campaign, 1865, Civil War

General Geary wrote, "And now we can truly say "Johnny comes marching home" in the language of the song. Thus you see I had the honor to participate in the last campaign of this ever memorable struggle…The glory and honor be to God, to whose great name we ascribe all praise." APGTW p. 242

April 28, 1865 The Second Division marched back to their previous camp near Raleigh.

April 29, 1865 The Second Division remained in camp near Raleigh and turned in all their ammunition except 25 rounds per man and loaded their wagons with supplies and forage and got ready to march to Washington D.C.

April 30, 1865 The Second Division marched second in line in the Corps following First Division, leaving at 7:30 a.m. They passed through Raleigh and crossed the Neuse River on a rickety bridge at the falls of Neuse Paper Mills and camped on Aligree's plantation, the bridge was repaired by the First Division but broke down before all the wagons of the Second Division could get over. The remaining wagons forded the river below the bridge and reached camp during the night. Major Spaulding reported, "Remained in camp until April 30, when, in compliance of orders from Gen. Barnum, the command left Raleigh, N. C., at 7 a.m.; crossed Crab Tree Creek and bivouacked on the bank of the Neuse at 9 p.m., the bridge having broken down, making it impossible for wagon trains to cross. This regiment was rear guard of the brigade." WOTR Vol. 47 p. 763

May 1, 1865 The Second Division marched in the lead and left at 5 a.m. taking the road toward Dickerson's Bridge. They crossed Cedar Creek on an excellent bridge at Long's Mill. Reaching the Tar River at 3 p.m., they found that Dickerson's Bridge had been washed away. The pioneers from the Second Brigade had been sent in advance and built a footbridge for infantry while the horses forded the stream. A mile further up the stream a pontoon bridge was constructed for the wagons to cross. They camped that night on the farm of B.A. Capehart having traveled 22 miles.

May 2, 1865 The Second Division was again in the lead and left at 5 a.m. for Williamsborough, by the way of Dimonds and Salem Crossroads. At Salem, they found the 14th Corps had left the road that was assigned to them and were marching on the road the 20th Corps was on. So the 20th Corps took a road to the right and then camped at 2 p.m. near Williamsborough having traveled 19 miles with little water.

May 3, 1865 The Second Division of the 20th Corps was in the front of the corps and they left at 4:30 a.m. passing through

Williamsborough and following the road toward Taylor's Ferry. Here both the 14th and the 20th Corps had to cross the Roanoke River, but it was too wide for two pontoon bridges. So the 20th Corps camped while the 14th crossed.

The Third Brigade moved out at 5 a.m. and marched 10 miles and then stopped at 10 a.m. near the Virginia state line to let the First and Third Divisions of the 20th Corps as well as the Fourteenth Corps to pass them.

May 4, 1865 The Second Division was in the rear of the corps, so they did not march until 6 a.m. They crossed Roanoke River at Taylor's Ferry on a pontoon bridge 385 yards long. After they crossed, they took the road to Saffold's Bridge over Meherrin River and camped four miles later having traveled a total distance of 22 miles. The Third Brigade resumed their march at 7 a.m. and crossed the Roanoke River at 8 a.m. and then camped for the night three miles from Saffold's Bridge on the Meherrin River.

May 5, 1865 Even though they were in the rear, the Second Division marched at 5:30 a.m. and crossed the Meherrin River at Saffold's Bridge and took a direct road for Blacks and Whites Station on South Side Railroad. They crossed Flat Rock Creek and camped at Barnes' Farm after traveling 21 miles.

May 6, 1865 The Second Division was again in the rear of march and again left early at 5 a.m. They crossed the Nottoway and Little Nottoway rivers on good bridges and passed Black and Whites stations and camped near Wellville Station after traveling fifteen hot miles with little water.

May 7, 1865 The Second Division was again in the rear of the corps. They left at 6 a.m. and went toward Kidd's Mill where they crossed the Appomattox River near Bevill's Bridge on pontoon bridges and then camped a mile beyond on Hawser's Farm having traveled that day 20 miles.

May 8, 1865 The Second Division was still in the rear of the corps and left at 6 a.m. and passed through Clover Hill Coal Mines, which were in full operation. Then they crossed Swift Creek and Falling Creek and camped there having traveled 21 dusty and dry miles under very hot weather.

May 9, 1865 The Second Division moved a few miles closer to Richmond and pitched camp in a large open field. They would stay there for a few days.

May 11, 1865 The Second Division resumed its march toward Washington. They were second in the marching order and left at 10 a.m. They passed through Manchester and Richmond and crossed the James River on a pontoon bridge and camped at Brook Creek four miles beyond the city on the Fredericksburg road having traveled 12 miles that day.

May 12, 1865 The Second Division was now second in line, and they marched at 7 a.m. It rained heavy the night before, and the streams were swollen and the whole area was swampy anyway, so it made for bad roads. They crossed Brook Creek, Chickahominy Creek and swamps and camped at Ashland on the railroad having marched 12 miles that day.

May 13, 1865 The Second Division was still second in line, but they left at 5 a.m. and taking the road toward Spotsylvania Courthouse crossed the South Anna and Little Rivers and then camped on the north bank of the Little River after having traveled 16.5 miles.

May 14, 1865 The Second Division marched in the same position in the corps and left at 5 a.m. They crossed the North Anna and Mat Rivers. The Third Brigade which contained the 102nd NY Veteran Volunteers and Sgt. Moses Whitbeck led the Division, and they went through Chilesburg and camped six miles from the Spotsylvania Courthouse at 5 p.m.

May 15, 1865 The Third Brigade marched at 5 a.m. and went through Spotsylvania and Chancellorsville battlegrounds where they had been in action two years before. They crossed the Rappahannock River at the United States Ford, as well as the Ta, Po, and Ny rivers. They stopped at 7 p.m. and camped on the north bank of the Rappahannock River having traveled 21 miles.

May 16 1865 The Second Division now marched in the front of the corps, left at 4:30 a.m., and headed toward Brentsville. They passed Hartwood Church and camped at Town Creek near Bristersburg. It was a very hot day and water was scarce and the roads dusty making for a tough march of 17 miles for the troops.

May 17, 1865 The Second Division was once again leading the corps and they left early. The weather was hot, as reported by General Geary, "Marched in advance of the corps at 4.30 a.m., crossed Cedar Run, and encamped at Brentsville. The weather grows more oppressively hot; distance 16 miles." WOTR Vol 47, pt. 1 p. 702

May 18, 1865 The Second Division was now in the rear of the corps and didn't march until about 10 a.m. They crossed Broad Run and Bull Run at Woodward's Ford, and camped within two miles of Fairfax Station. It was a very hot day until a thunderstorm arose about 4 p.m.

The Third Brigade started at 9:30 a.m. toward Fairfax Station, Virginia. This was a place they knew well. But first they had to withstand the severe rainstorm when they were about a mile from Fairfax and they had to stop for the night. They still managed 14 miles that day on bad, hilly roads.

May 19, 1865 The Second Division was still in the rear of the corps and marched around 6 a.m. They passed Fairfax Station and followed a small road which entered the Alexandria turnpike between Fairfax Courthouse and Annandale. Then they took the turnpike and marched until they got to within 3.5 miles from Alexandria where they camped on Gregory's Farm.

The Third Brigade resumed their march toward Fairfax Station at 6 a.m. After they passed through the town, they took the Warrenton and Alexandria pike and arrived at Cloud's Mills at 5 p.m., having traveled about 18 miles. They camped on the land between the Warrenton and Leesburg pikes.

May 20, 1865 The Second Division stayed in camp at Gregory's Farm, ordering supplies for the command and getting cleaned up for the Grand Review in Washington D.C. on May 24. Colonel Chatfield reported, "In relation to the operations of this regiment since leaving Richmond, I have the honor to report that on the 11th day of May, 1865, it left camp near Manchester, Va., and commenced its march toward Washington, in pursuance of orders received from army headquarters, passing through Richmond the same day. Marched via Chilesburg, Brentsville, and Spotsylvania Court-House, and arrived near Alexandria on the 19th day of May, where it encamped. On the 24th of May this regiment broke camp, marched to and through Washington, passing in review at the latter place, and arriving at our present encampment the same day. During the march good order and discipline was preserved throughout the command and the property of private citizens at all times respected." WOTR Vol 47. p. 762

May 24, 1865 The 20th Corps broke camp early and headed for Washington. They moved to Long Bridge and then passed in review up Pennsylvania Avenue and through the city, then camped in the vicinity of Bladensburg, four miles from Washington. D.C. Although the Administration quickly scaled down the size of the Army, General Geary and the Second Division remained in the service until **July 19, 1865**.

July 1, 1865 Moses Whitbeck was promoted to sergeant-major.

July 21, 1865 Moses Whitbeck was mustered out with the regiment at Alexandria, Virginia.

Aftermath and Legacy

The following is taken from The Union army: a history of military affairs in the loyal states, 1861-65 -- records of the regiments in the Union army -- cyclopedia of battles -- memoirs of commanders and soldiers. Madison, WI: Federal Pub. Co., 1908. volume II.

During its long and honorable service the 102nd buried its dead in seven states, and participated in over 40 battles and minor engagements. It participated in many a famous charge, one of the most gallant being at Lookout mountain, where the regiment, as part of Ireland's brigade, struck the enemy on the flank and drove him in confusion from the field. It belonged to the gallant White Star division, commanded by Gen. Geary, who complimented the regiment as follows: "It may safely be asserted that no organization in the army has a prouder record, or has passed through more arduous, varied and bloody campaigns." The loss of the regiment during service was 7 officers and 67 men killed and mortally wounded; 82 men died of disease, accident, etc., a total of 7 officers and 149 enlisted men. The gallant Maj. Elliott was killed in action at Lookout mountain.

One Hundred and Second Infantry.—Cols., Thomas B. Van Buren, James C. Lane, Herbert Hammerstien, Harvey S. Chatfield; Lieut-Cols., William B. Hayward, James C. Lane, Harvey S. Chat-field, Oscar J. Spaulding; Majs., James C. Lane, F. Eugene Trotter, Gilbert M. Elliott, Lewis R.. Stegman, Oscar J. Spaulding, Reuben H. Wilber.

This regiment, known as the Van Buren light infantry, was principally recruited at New York city, and was formed by the consolidation of the Von Beck rifles under Col. R. H. Shannon, and part of the McClellan infantry under Col. S. Levy, with Col. Van Buren's command. The organization was completed later by the addition of two companies from the 78th Cameron Highlanders and Co. A, 12th militia, and was mustered into the U, S. service from Nov., 1861, to April, 1862.

In July, 1864, its ranks were filled by the transfer of the officers and men of the 78th N. Y. infantry. On the expiration of its term of service the original members (except veterans) were mustered out, and the regiment, composed of veterans and recruits continued in service. Early in June, 1865, it received by transfer the remaining men of the 119th, 154th, 137th, 149th, 134th, and 184th N. Y. Vols.

The regiment, eight companies, left the state on March 10, 1862, followed by Cos. I and K on April 7. Assigned to the 2nd brigade, 2nd division, 2nd corps, Army of Virginia, it fought its first severe engagement at Cedar mountain, where its loss was 115 killed, wounded and missing.

The regiment then moved with its corps to the support of Pope, fought at the second battle of Bull Run, and went into position at Chantilly, but was not engaged. In the same brigade and division, 12th corps, it was actively engaged at Antietam, losing 37 killed, wounded and missing, and. was then successively engaged in the minor actions at Lovettsville, Ripon, Hillsboro, Winchester, Wolf Run shoal, and Fairfax Station, going into winter quarters at Stafford Court House.

At the battle of Chancellorsville the l02nd, which fought in Geary's division of the 12th corps, lost 90 killed, wounded and missing. It was heavily engaged with the "White Star" division at Gettysburg, where its total loss was 29. It followed with its corps in pursuit of Lee's fleeing army, being engaged at Ellis' ford and Stevensburg, and in the latter part of September moved with the corps to Tennessee to reinforce Gen. Rosecrans. It engaged in the midnight battle of Wauhatchie; then started on the Chattanooga and Rossville campaign, fighting the famous "Battle above the clouds" on Lookout mountain, where the division led the advance; then fought at Missionary ridge and Ring-gold gap, its loss in the campaign being 14 killed, wounded and missing.

In the same brigade and division, 20th corps, the 102nd was with Gen. Sherman all through his Atlanta campaign, fighting at. Villariow, Mill Creek Gap, Resaca,. Calhoun, Cassville, Dallas, Acworth, Kennesaw Mountain, Chattahoochee River, Peachtree Creek, where its losses amounted to 53 in killed, wounded and missing. It moved in November with Sherman's army on the march to the sea, shared in the siege of Savannah, its active service closing with the campaign in the Carolinas, during which it was engaged at. Wadesboro, Averasboro, Bentonville, Goldsboro, Raleigh, and Bennett's house, losing 18 killed, wounded and missing during this final campaign.

It was mustered out under Col. Chatfield, July 21, 1865, at Alexandria, Va.

This is the end of the incredible story of the 102nd NY regiment in the Civil war. I hope you enjoyed reading the story.

102nd NY Officers and Commanders

Some of the commanders and officers of the 102nd and what happened to them.

Thomas B. Van Buren, Colonel. He was a lawyer and member of the New York state assembly, and son of Peter Van Buren and Mary Brodhead. During the Civil War he recruited and was the Colonel of the 102nd New York State Volunteers. He participated in the battles of Antietam and Cedar Mountain, (but not Second Bull Run as some have suggested or

"Battle" Mountain which is incorrect as they meant to say Cedar Mountain).

He left the service due to poor health. Even though he helped build the regiment, he served less than a year. He was brevetted Brigadier General by President Lincoln in 1865. After the war he became US Consul General at Kanagawa Japan from 1874 to 1885 and author of "Labor and Porcelain in Japan." The book's purpose was to inform the U.S. public of Japan's industry in order to stimulate U.S. imports from Japan and it contained very early and scarce photographic images of the Emperor of Japan and his wife. He died Oct 13 1889 (age 65) in San Francisco. Find a Grave.com: Thomas Brodhead Van Buren

William B. Hayward, Lieutenant Colonel. He was originally the commander of the 60th New York Volunteers, but the regiment's company commanders petitioned for Colonel Hayward's removal, so he was sent to the 102nd NY. Although he was second in Command when the regiment left New York, he was discharged July 16, 1862.

James C. Lane Major. He was born in July 23, 1823 in New York City. He graduated Green Mountain College, Class of 1841, and became an architect and civil engineer. He worked on construction of the Illinois Central Railroad, the US Coast Survey, and on mining surveys of Cuba and Puerto Rico.

When the 102nd New York State Volunteers were organized in March 1862, Lane was appointed Major. He was promoted to Lieutenant Colonel just before the Maryland (Antietam) Campaign. The 102nd NY then fought at Cedar Mountain and Antietam under Lt Col Lane. He was promoted to Colonel on December 14, 1862. He was wounded July 2nd 1863 on Culp's Hill at Gettysburg.

In October 1863 the 12th/20th Corps was sent to the western theater with Colonel Lane rejoining his regiment in time to lead the attack on Lookout Mountain during the so called "battle above the clouds." During the Atlanta campaign in which the durable marching 102nd saw its share of combat, Colonel Lane mustered out on July 12, 1864, still suffering the effects of his Gettysburg wound.

After the War he continued to work as a mineralogist, doing surveys in California, Arizona, and Nevada. He was chief engineer of the New York, Woodhaven and Rockaway Railroads and later designed parks in Westchester County. He died December 12, 1888 (age 65) in New York City.

Lewis R. Stegman Captain – He had been a surveyor and began a law practice in New York City by 1861. On 30 September of that year, then 22 years old, he enlisted in New York City, helped recruit men for the Company, and mustered as Captain, Company E.

He was wounded in the regiment's first battle on August 9, 1862, the Battle of Cedar Mountain. As the leftmost unit during the Union advance on the southern portion of the battlefield, the 102nd New York was exposed to Confederate artillery fire. A shell exploded near Stegman and a fragment struck him in the head. Nevertheless, Stegman refused to sit out the Battle of Antietam the following month and went into action with a bandage on his head. He remained in command of Company E throughout the rest of 1862 and into 1863, leading it in the battles of Chancellorsville and Gettysburg. In the latter engagement, Stegman took command of the regiment when Col. James C. Lane was wounded. On November 24, 1863, Stegman led his company during the successful assault on Lookout Mountain. Three days later, at the Battle of Ringgold Gap, he distinguished himself by leading 40 men to prevent the Confederates from burning a bridge as the enemy retreated. On June 16, 1864, Stegman was shot through the thigh while commanding a skirmish line during the battle of Pine Knob. Because of his wound he had to resign from the 102nd NY. After he healed, in October 1864, he became major of the 1st United States Veteran Volunteers serving in the Shenandoah Valley and Loudoun County, Virginia, but saw no major action. After the war ended in 1865, he applied for a commission in the regular army but was turned down. Like many veteran officers, Stegman received brevet promotions for gallantry and went home a colonel of U.S. Volunteers.

After the Civil War, Stegman returned to New York and made his home in Brooklyn, working in a number of professions, most often as a journalist. In 1876 he became under-Sheriff of Kings County, New York, thus beginning a career in public service that included serving in the New York State Assembly (Kings Co., 6th D.) in 1879 and as Sheriff of Kings County beginning in 1881. His election as Sheriff was noteworthy as he was elected as a Republican in a largely Democratic district. In 1886, he was indicted for stealing from an estate. The outcome is unknown, but fallout from the case was heard in civil suits and by the NY Supreme Court to at least 1896. Antietam on the Web

During his final years, Stegman was a prominent member of the New York Monuments Commission, working to commemorate the sacrifices of New York soldiers during the Civil War. He wrote numerous historical essays and biographies that are contained in a book he authored with Joseph A. Joel. He wrote a stirring account of the battle of Gettysburg and the fight on Culps Hill that involved Greene's Brigade and the

102nd NY. He was a featured speaker at many veterans' events including monument dedications at Gettysburg and Antietam. Stegman became chairman of the commission in 1912 after a financial scandal ousted chairman Daniel E. Sickles, and presided over the dedication of the State monument at Antietam on September 17, 1920. Lewis R. Stegman died on October 7, 1923, at age 84, in Brooklyn, New York.

Herbert von Hammerstein Colonel. Before the Civil War he was an officer in the Austrian Calvary. He was originally a Captain in the 8th New York but then was discharged and became a Major in General McClellan's staff. He remained there until June of 1863 when he was transferred to the 78th New York as Lt. Colonel. He commanded the 78th NY during the Battle of Gettysburg. He came over to the 102nd NY when they consolidated units. He was promoted Colonel to command the 102nd NY when Colonel Lane left because of his wounds. After the War he remained in the service until discharged in 1867 for disability.

Harvey S. Chatfield Colonel He served in the 78th NY until July 12, 1864, at which time he was promoted to Lieutenant-Colonel of the 102nd NY. On June 4, 1865, he was promoted to Colonel, and served in that rank until the regi-ment was mustered out, July 21, 1865, at Alexandria, Virginia.

Oscar J. Spaulding Major - commanded during the Carolinas Campaign, he joined the 102nd from the 78th when they consolidated.

Arthur Cavanagh. Mortally wounded in action at Cedar Mountain on August 9th 1862, died August 28, 1862. He was a Captain in charge of Company G, which was Moses' company, but I have not found any more details about him.

Gilbert M Elliott, Major – Was 1st Lieutenant in October

1861, rose to Captain October 29, 1862, became a Major June 19, 1863. Killed in action Lookout Mountain, November 24, 1863.

M. Eugene Cornell, Captain Company D, killed at Antietam September 17, 1862. He was shot in the head while bringing his men into line of battle.

Robert Avery Captain He was wounded May 3, 1863 at Chancellorsville and then again November 24, 1863 at Lookout Mountain. He was discharged June 27, 1864 for disabilities from his wounds. He do know that General Geary recommended him for promotion to Colonel of the 102nd NY on December 24, 1862 (Letter from General Geary to Edwin D. Morgan, the Governor of New York.)

Julius Spring Captain. He was killed in action August 9, 1862 at the Battle of Cedar Mountain.

John Meade Captain. He rose from First Lieutenant to Captain. He was killed in action July 3rd, 1863 at Gettysburg.

William T. Forbes First Lieutenant He was 26 years old when he enlisted at Brooklyn, N.Y., on 11/22/61. He mustered into Company G, Moses Whitbeck's unit. He rose from First Lieutenant to Captain in November of 1862. During the Gettysburg campaign, he was Acting Assistant Inspector General, 12th Corps, Then on November 17, 1864 he was commissioned into the U.S. Volunteers as major in the Adjutant Generals Department.

J. Virgil Upham, Adjutant. He rose from being a First Sargent in the 78th to First Lieutenant, Company C. He then became Adjutant when he joined the 102nd. He was killed in action July 2, 1863 at Gettysburg.

Francis Bacon Second Lieutenant. He was killed May 3, 1863 at the Battle of Chancellorsville.

Letter from Frank L. Whitbeck
(Grandson of Moses)

SERGEANT MAJOR MOSES WHITBECK
From Ulster County, New York, 102d N.Y. Volunteers
WAR BETWEEN THE STATES – USA

On November 13, 1861, Moses Whitbeck, a farmer in Ulster County, New York, was enrolled in Co. F, 1st Regiment, Von Beck Rifles, which unit was organized in the City of Kingston at the beginning of The American Civil War. This organization became Company "G" of the 102d N.Y. Volunteers in January or February, 1862. He remained with this group until his discharge from active duty on July 21, 1865, at a camp of the The Army of the United States near Alexandria, Virginia.

Whitbeck was born in Ulster County, N.Y., on June 11, 1842. He was the first-born child of Jacob and Margaret Pattison Whitbeck, who were married in the Dutch Reformed Church in Hurley, N.Y., the previous year. Thus Moses Whitbeck was 19 years, 5 months and two days old when he enlisted. He was discharged after serving the Federal Army a total of 3 years, 8 months, and 8 days. During this period he served successively as private, corporal, sergeant, and sergeant major in campaigns in Virginia, the Carolinas, Tennessee, and Georgia until he was severely wounded near Dallas, Georgia, on May 29, 1864. He overcame a serious head wound (right side), and was hospitalized at Chattanooga, Tennessee, before being transferred to Alexandria, Virginia.

The foregoing information is confirmed by family biblical records and letters, as well as by certified copies of Moses Whitbeck's original enlistment, his reenlistment after three years, Company "G" morning reports, casualty lists, hospital records, honorable discharge from the Army, application for pension, certification by his second wife, Frona Carley Whitbeck, and their sons (Frank, Roy, and Neal), death certificates, and finally Frona Whitbeck's widow's pension file.

However, there is still a discrepancy or omission I have not been able to clear up. After personal visits to Chickamauga, Kennesaw Mountain, Pickett's Mill, New Hope Church, and Dallas battlefields and monuments in Georgia, I have not uncovered any positive record the Co. "G" of the 102d N.Y. Volunteers as a separate unit of infantry was present during these most extensive and frightening engagements. In one report, Sergeant Major Moses Whitbeck was shown to be attached to the "2nd Division, 20th Corps Infantry." The Atlanta Constitution, one of the South's leading newspapers, on Sunday, July 17, 1994, published full information on the relentless Union drive toward their strategic goal – Atlanta. The paper reported that on May 28, 1864, there were 380 Union and 600 Confederate casualties at Dallas (about 40 miles west/northwest of Atlanta). Moses Whitbeck suffered "a severe gunshot wound to the right side of his head while bearing a rifle" the very next day. I plan to continue searching the known records of Co. "G", 102d NY Volunteers, through libraries and other means. I suspect that I shall find that Co. "G" was assigned to the Army of the Cumberland under the command of General Thomas, which Army was credited with the successful breakthrough at Dallas in its drive toward Atlanta.

Incidentally, Moses Whitbeck was paid a small bounty on completion of this three years' service. He was also paid a small bounty when he was finally honorably discharged. Moses Whitbeck died at Shawnee, Oklahoma, on March 31, 1906, from pneumonia; he was buried in the southwest section of Fairview Cemetery. A daughter of his first marriage, Minnie Rose Whitbeck (who never married), is also buried in this same plot. Both Frona Whitbeck and Mary Elthea Vanwagenen Whitbeck, are buried in Fredonia Cemetery (Wilson County), Kansas.*

*Subsequent to his return to Ulster County, NY, Moses Whitbeck married Mary Elthea Vanwagenen. They had three children, all presumably born in Ulster County: Harry b. August 18, 1869; Minnie Rose b. October 5, 1870; and Laura b. March 7, 1876. For some reason the opening West lured Moses and family to Wilson County, Kansas (Fredonia). Mary Elthea died unfortunately on August 31, 1877, leaving Moses with a family to raise alone. On December 12, 1882, he married Frona Carley at Altoona, Kansas, Wilson County, she was then age 29, he 40. Frona took care of the children as her very own and won lifetime support and admiration from them until her death. She had four children of her own: Mignonette b. March 22, 1884, d. March 26, 1884; Frank b. December 6, 1885; Roy b. September 22, 1887; and Neal b. March 21, 1890. Frona's pension as a civil war widow began at $8.00 per month. Before she died the pension had been increased to $24.00 per month. Leaving Kansas this family pioneered the opening of Oklahoma Territory.

Frank L. Whitbeck (grandson of Moses)
Littlerock, Arkansas

Bibliography

A Politician Goes to War, by The Pennsylvania State University Press. This book contains the letters That General John W. Geary wrote to his wife. These quotations are listed as APGTW, with the page number the quote is from listed.

From the Canon's Mouth, by Wayne State University Press and the Detroit Historical Society. This book contains the letters that General Alpheus Williams wrote to his daughters. These quotations are listed as FTCM, with the page number the quote is from listed.

A Perfect Storm of Lead, by R. L. Murray. These citations are listed as APSL, with the page number the quote is from listed.

The War of the Rebellion: A Compilation of the Official Records of the Union and Confederate Armies. Some researchers refer to them as the O.R. In this book they will be referred to as WOTR. The citation are listed as WOTR with the Volume number, Part number if there is more than one, and page the citation if from listed.

New York State Military Museum Official Unit roster, 102nd New York Volunteer Regiment.

Wikipedia: Irvin McDowell; Mud March; Chattanooga campaign; Battle of Missionary Ridge; Battle of Atlanta; Atlanta campaign; The Patrick R. Cleburne Confederate Cemetery; Goldsboro, North Carolina

Battlefield.org: 10 Facts Harpers Ferry

Encyclopedia Virginia.org : Winchester during the Civil War, Gordonsville during the Civil War; American Battlefield Trust: Cedar Mountain; telegram-from-george-b-mcclellan-to-abraham-lincoln-september-13-1862

The Virginia Department of Historic Resources: Fairfax Court House 1861-1865

Shades of Blue and Gray: Georgetown and the Civil War | Georgetown University Library. https://library.georgetown.edu/exhibition/shades-blue-and-gray-georgetown-and-civil-war

History.com: This-day-in-history/union-troops-discover-rebels-antietam-battle-plan; Sherman's March to the Sea; Union army sacks Columbia; South Carolina: blue-gray-war-soldiers-letters

The American Battlefield Trust: The Maryland Campaign of 1862; Kelly's Ford Battle Facts and Summary; Chancellorsville Articles; The Etowah River; Cobb County in the Atlanta Campaign; Pickett's Mill Battle Facts and Summary; Dallas Battle Facts and Summary; Bennett Place Surrender

Harpers Ferry National Historical Park (U.S. National Park Service): Loudoun Heights

Bolivar Heights (U.S. National Park Service).

https://www.nps.gov/places/000/bolivar-heights.html

db0nus869y26v.cloudfront.net/en/George_S._Greene

Seeing the Civil War -Army of the Potomac

Infogalactic: John S. Mosby

The American Presidency Project - Recalling Soldiers to Their Regiments

New York in the War of the Rebellion

Historical Marker Project. Hartwood Presbyterian Church - Fredericksburg - VA - US

The Civil War Months: Hooker Moves in Earnest

Academic.com: Battle of Chancellorsville.

Cloudfront.net: Battle of Chancellorsville; Chattanooga Campaign

Historynet.com: Battle of Chancellorsville; All or Nothin': The surrender Sherman and Johnston crafted at Bennett Place.

History Maps.com: Battle of Chancellorsville

Thomaslegion.net: American civil War: Chancellorsville Campaign

Seeing the Civil War: Brandy Station VA

Culp's Hill at Gettysburg, John M. Archer
MyCivilWar.com: The Battle of Rocky Face Ridge; The Battle
of Adairsville.

New World Encyclopedia: Atlanta Campaign

Atlanta Journal Constitution: May 22, 2014

AmericanHistoryCentral.com: Battle of Kolb's Farm, 1864;
Carolinas Campaign; Carolinas Campaign, 1865, Civil War

Letter of William T. Sherman to Henry Halleck, December
24, 1864

William T. Sherman, Message to William J. Hardee,
December 17, 1864, recorded in his memoirs.

Dead Confederates.com: The-burning-of-columbia-february-
17-1865

North Carolina History Project: Carolinas Campaign (January
1865-April 1865)

Civil War Era NC: Fayetteville, Union Accounts

Ohio Civil War Central.com: Battle of Averasboro

Emerging Civil War.com: The Last Charge of the Army of
Tennessee, original quote can be found: Mark Bradley,

Bentonville (Campbell, CA: Savas Beatie, 1996), p. 204; Last Stand in the Carolinas: The Battle of Bentonville. p. 403-404

Pillar to Post: RETRO FILES / CIVIL WAR AT THE BITTER END.

Heritage Post.org: The Battle of Bentonville

Christopher Klein: 10-things-you-may-not-know-about-the-lincoln-assassination.

ABOUT THE AUTHOR

CHRIS WHITBECK is an Air Force veteran who worked in the aerospace sector for 15 years as an RF Technician and has worked with the Washoe County School District for 16 years teaching computers and sports. He officiates high school and youth sports and serves as secretary of the Northern Nevada Football Officials Association and the Northern Nevada Lacrosse Officials Association. Chris and his wife Jill have four children and reside in Reno. He has studied history all his life and is researching a book about his great-great-grandfather Jonathan Miller and the 93rd Illinois Volunteers during the Civil War. "Moses Whitbeck and the 102nd NY Volunteers During the War of Rebellion" is his first book.

Made in the USA
Las Vegas, NV
25 January 2024

84838834R00167